Digital Technologies and African Societies

*To my wife Fatou and my children Eva, Nelly, Marie and Tite*
*and all those who have dedicated their time to my scientific productions*

Julien Atchoua

*To my partner Laetitia and my children Jessica and Ethan*
*and all those who have contributed to the realization of this book*

Jean-Jacques Maomra Bogui

*To my wife Faby and my son Eja*
*and to those who derive their happiness from serving others*

Saikou Diallo

**Africa Digital Age Set**

coordinated by
Jean-Paul Bourrières

Volume 1

# Digital Technologies and African Societies

*Challenges and Opportunities*

*Edited by*

Julien Atchoua
Jean-Jacques Bogui
Saikou Diallo

ISTE

WILEY

First published 2020 in Great Britain and the United States by ISTE Ltd and John Wiley & Sons, Inc.



ISTE Ltd
27-37 St George's Road
London SW19 4EU
UK

www.iste.co.uk

John Wiley & Sons, Inc.
111 River Street
Hoboken, NJ 07030
USA

www.wiley.com



Library of Congress Control Number: 2020935386

British Library Cataloguing-in-Publication Data
A CIP record for this book is available from the British Library
ISBN 978-1-78630-451-3

MIX
Paper from responsible sources
FSC www.fsc.org FSC® C013604

# Contents

# Foreword

In a series of conferences entitled *Penser global* (thinking global), Edgar Morin proposed combining "development" and "envelopment": "All that is development in the classical sense of the term must be accompanied by what surrounds it, i.e. communities, fraternities."[1] This book, edited by Julien Atchoua, Jean-Jacques Bogui and Saikou Diallo which I have the honor to write the foreword for, provides a very insightful review of the link between development and envelopment. Rooted in the multiple realities of African societies and their relationship to ICTs (Information and Communication Technologies), the various texts in this book provide valuable insights into the diversity of situations and the dynamism of socio-cultural appropriation of digital technology.

Before focusing on everything that makes a difference, one thing must be noted: the omnipresence of technologies in the many sectors of our societies, and beyond, in the very weaving of our lives, even if this takes place to different degrees. Digital technology is based on a process of discretization of data and declined through multiple techniques, at the forefront of which are the Internet and the Web. It deploys a hegemonic presence, whose reality is no longer discussed today. If the history of technology allows us to situate the digital "phenomenon" in relation to other major breakthroughs, such as printing[2], the fact remains that certain dimensions specific to digital technology give it a quite exceptional character. The massive convergence of

1 Morin, E. (2015). *Penser global. L'humain et son univers.* Robert Laffont/Éditions de la Maison des sciences de l'homme, Paris, 82.

2 Eisenstein, E.L. (1991). *La révolution de l'imprimerie: à l'aube de l'Europe moderne.* La Découverte, Paris.

all types of data on media and in formats that allow automated computational processing profoundly challenges not only the traditional frameworks for the exercise of thought or the production of knowledge, but also our perception of the world. Digital tools help shape our thoughts, representations, cultures and societies. Digital technology is becoming a real "environment"[3], an ecosystem in which we act, interact, evolve and live. Thus, the conditions for a "digital humanism" are developed, because "the current technology, in its global dimension, is a culture in the sense that it sets up a new context on a global scale [...]"[4].

Once these major precepts of digital technology have been established, it is essential to add a few nuances. Not all countries, societies and individuals are equal in terms of equipment, access and skills. Reiterating the false evidence of digital technology and its associated technologies still seems necessary at a time of supremacy that tends to obscure many persistent inequalities. The concept of the "digital divide" has been around for a long time and remains, in its own way, a relevant lever for maintaining vigilance with regard to certain implicit factors related to the use of digital technology. The political–industrial policies and initiatives that multiplied in the 2000s put the issue of equipment and materials at the top of their agenda. In order to fill a gap in equipment (mobile phones, computers, etc.) and access (mainly to the Internet), these policies and initiatives have facilitated inclusion in a connected society. In this context, it is assumed that the reduction of technological inequality (concerning equipment, access or skills) will inevitably require the reduction of other inequalities (social, cultural, cognitive, to name but a few).

Several chapters in this book testify to this greater accessibility, whether with distance learning or through the role of mobile telephony in financial transactions. The commoditization of mobile phones in African countries thus makes it possible to reconfigure economic dynamics (from a centralized model to a decentralized model) while embracing social and cultural logics, whose characteristic examples can be seen through the circulation of money in connection with migration flows or within a country. This flexibility provided by the widespread presence of connected technologies makes it possible in this context to promote inclusion (societal and financial) and

3 Bachimont, B. (2015). Le numérique comme milieu: enjeux épistémologiques et phénoménologiques. *Interfaces numériques*, 4(3), 387–404.
4 Doueihi, M. (2011). *Pour un humanisme numérique*. Le Seuil, Paris.

reduce significant territorial inequalities. Health is another major area where the use of technologies, and here too mobile telephony (in a mobile health[5] approach), is a valuable means of reducing inequalities, whether they be geographical or professional. Gilbert Toppé's text on telemedicine and the Genesis Telecare platform in Cameroon (Chapter 11) clearly shows the possibilities of developing access to quality care.

Nevertheless, while there are still strong questions about equipment and access (such as the problem of electricity in Africa), and while this is an essential prerequisite for the development of uses, it is not enough. On the one hand, it would be illusory to hope for equity through a utopian alignment of equipment indicators. Moreover, when the use of digital technology and the multiple devices that accompany it is reduced to their purely technical dimension, it results in denying the sources of uncertainty, the margins of interpretation, the socio-cultural dimensions or even the heterogeneous forms of otherness. Thus, other dimensions must be considered in interpreting the possible "gaps" in relation to digital technology, such as the deficiencies in the mastery of knowledge and skills related to an information and digital culture or even the know-how in terms of methods. Beyond a reductive vision of a "gap" to be filled in terms of equipment, it is therefore necessary to think about the differences that underlie singularities, whether at the individual level or at the level of the collectives, people and communities that make up the essence of this digital "multiversity".

Another approach to the false evidence of digital technology is particularly obvious with digital social networks. Several chapters in this book show the very positive role that these social networks, including applications such as WhatsApp or the iconic Facebook platform, have had in the mobilization and participation of citizens in the processes of democratic vigilance and public action. The case of the *Balai Citoyen* (Citizen's Broom) in Burkina Faso is exemplary in this respect. Since the first Arab revolutions of 2011, in Tunisia or Egypt, examples have multiplied throughout the world, showing the potential for increased mobilization through these social media, as well as relaying demonstrations and actions in the physical world. This reveals the extent to which traditional institutions (political, media) can in a way be overwhelmed. However, the weight of the technical configuration of

5 Reference can be made to the reports and work of the World Health Organization, including the article: mHealth – Use of appropriate digital technologies for public health [Online]. Available at: http://apps.who.int/gb/ebwha/pdf_files/WHA71/A71_20-en.pdf.

the device and the nature of the algorithm should not be underestimated in managing affiliations and connections authorized, and therefore oriented, by the social network platform. Thus, it was shown[6] that the success of the Yellow Vests movement in France at the end of 2018 and the beginning of 2019 was correlated with the change in Facebook's algorithm in January 2018, favoring the refocusing of information on local communities or communities of similar interests, which has encouraged a viral spread of mobilization throughout France. The system proposed by Facebook thus emerges as an "algorithmic facilitator" of politics, bypassing institutional obstacles to the democratic and reactive circulation of a citizen's voice. But this is inevitably accompanied by notorious perverse effects. The Cambridge Analytica scandal, the manipulation of voters via Facebook or Twitter during American or English election campaigns, and more generally, the place of fake news in the digital information panorama, illustrate this other obscure side of the power of digital social networks.

Through these few milestones, we can clearly see the ambivalence of digital technology and its technologies, between curative and toxic effects, which can be described as "pharmacological" if we take up the terminology and reading proposed by Bernard Stiegler[7]. This intrinsically pharmacological nature of digital technology is pertinently illustrated by the various texts of this book, between the "positive" uses of technologies for financial flows, training, access to healthcare or even political and citizen mobilization and their "deviant" uses such as those practiced by grazers in Côte d'Ivoire, with forms of know-how transmission and professionalization of cybercriminal activity, as well as that of scams on the Internet. Opportunities and risks therefore appear to be consubstantial with digital technology. From the birth of the Internet in the 1960s and the 1970s, between military and militant roots, to the arrival of the Web in the early 1990s and up to today, the human and social adventure of digital technologies is characterized by a profusion of initiatives, successes and major upheavals that constantly oscillate between gains in individual freedoms, the production of a collective voice (via digital social networks),

6 See, for example, Bornstein, R. (2019). En immersion numérique avec les Gilets jaunes. *Le Débat*, 2, 38–51.

7 Tout objet technique est pharmacologique: il est à la fois poison et remède [Online]. *Vocabulaire Ars Industrialis*. Available at: http://www.arsindustrialis.org/pharmakon.

new polarizations (particularly economic)[8] and poorly transparent calculation logics (governance by algorithms)[9].

At the origin of these crucial reconfigurations supported by digital technologies, three factors of change can be easily identified: spatial disruption, temporal disruption and ubiquity. While changes in space and time are part of the trajectory of a history of the information and communication media, they are currently taking place on a globalized scale and in an almost instantaneous manner that has no precedent. Ubiquity and access to the masses (singular or collective) represents a more unprecedented factor, with many effects whose characteristics are mentioned in this foreword and detailed in a number of chapters within this book. To these three factors, it seems appropriate to add a fourth one, which opens up perspectives of which we can only see the beginnings: the massive production of data, correlated with automated and often predictive processing. While most of the current volume of data is still of human origin (traces of web browsing, data from social networks), very soon, the widespread presence of passive sensors, coupled with the Internet of Things and artificial intelligence-type processing, will significantly increase the role and impact of these data in different sectors of society and human activity. Some of the major sectors that will be impacted by digital technology in the near future[10] include industry and services, education and training, culture and media, health, not to mention all the issues related to the environment and ecological transition.

The stakes are immense, the challenges are major and the future is always uncertain. Resituating these key issues in the light of the problems, uses and social innovations at work on the African continent, as this book does, is fundamental. Because the dynamics that are taking place on this continent are still new and show both patterns of use, they represent a certain universalism of digital technology and culturally situated appropriations inscribed in the matrices of societies and cultures. "Neither globalization nor digitization can, it seems, thwart the singular aspiration within a small

8 Just look at the current economic weight of the GAFAs (Google, Amazon, Facebook, Apple), and their Chinese equivalent, the BATXs (Baidu, Alibaba, Tencent, Xiaomi).

9 See, in particular, Rouvroy, A., Berns, T. (2013). Gouvernementalité algorithmique et perspectives d'émancipation. Le disparate comme condition d'individuation par la relation?. *Réseaux*, 1(177), 163–196.

10 Based on Cap Digital (2018). Cartographie des tendances 2018–2019 [Online]. Available at: http://www.capdigital.com/2018/05/15/cartographie-des-tendances-2018-2019/.

defined group for difference."[11] At the heart of these techniques and in connection with forms of injunction generalized to digital[12] participation, it is indeed, and always, the place of the human whose needs, aspirations, desires, dreams and singularities must be constantly reaffirmed. As such, while the inclusive potential of technologies seems undeniable, it is important to constantly think about the diversity of individual and collective situations. The always unique combination, of multiple heterogeneities, contributes to creating not only balances but also permanent imbalances between the individual, the social and the digital, and keys to access and differentiated actions to/from this digital "environment". Reality is always more complex – and unexpected – than its reduction to a single horizon. It is this openness that the various chapters of this book provide us.

Nathalie PINÈDE
Lecturer – HDR
Information and Communication Sciences
Institute of Technology
Bordeaux Montaigne University

11 Laulan, A. (2018). Diversité culturelle et mondialisation. *Hermès, La Revue*, 80(1), 168–174.
12 Proulx, S. (2017). L'injonction à participer au monde numérique. *Communiquer*, 20, 15–27.

# Introduction

## I.1. Problems with ICTs in development projects and activities in Africa

Like many other studies on specific sectors of development in Africa, this book, which focuses on the theme of digital technologies and African societies, is a set of reflections on the place of digital technologies in development projects and activities on the African continent. These reflections are based on the lines of research related to ICT (Information and Communication Technology) and human development, the digital economy, cybercrime, digital identification and monitoring, social media and public space, and ICT and the modernization of higher education in Africa.

Indeed, the integration and use of information and communication technologies (ICTs) in African countries are increasingly visible in various sectors of activity (banking, education, trade, etc.) despite a digital divide that is still relevant today. ICT has become a major sector in the recent growth of a new informal economy in African cities (Chéneau-Loquay 2008). The role of ICT has been at the heart of various international meetings. A generally positive, and even at times utopian, discourse is generally heard on the contribution of digital technologies to the development of African states. The adoption or appropriation of digital technologies by Africans is presented in many speeches by politicians or institutions involved in international cooperation and development as an important issue for the continent's development.

---

Introduction written by Julien ATCHOUA, Jean-Jacques BOGUI and Saikou DIALLO.

According to the UNDP (2001):

> "ICTs provide developing countries with an unprecedented opportunity to achieve vital development goals, such as poverty reduction, basic health care and education, much more effectively than ever before."

It is also in this same perspective that the economic successes generated by ICTs in Africa are very often highlighted in the speeches of politicians, academics and professionals working for international organizations specialized in development. Some authors consider them as an opportunity for the African continent (Bonjawo 2002).

The enthusiasm for mobile telephony among the African population is particularly remarkable. In 2004, UIT stated that Africa was the fastest growing mobile market in the world. In 2012, mobile companies contributed $21 billion to the government funds in the region. The contribution of the sector is expected to double by 2020 to 6.6 million employees and $42 billion in government revenue.

Donner (2006) studied the effect of mobile phones on small businesses in Rwanda, and concluded that the use of this technical tool has helped develop the networks of local entrepreneurs. According to Acker (2008), the use of mobile phones is transforming rural markets and increasing business opportunities for farmers in Niger. For Tall (2004), the mobile phone has promoted the reconnection of the extended family, particularly in rural Senegal. Exchanges – especially financial – between remote members and those who have remained in the country have become more numerous and more regular (Alzouma 2008).

The African continent is also experiencing significant growth in m-banking and m-payment services, but these services on the continent do not have the same meaning as in industrialized countries where they appear as additional services to the bank. In African cities, they appear as an alternative to traditional banking services that are inaccessible to the majority of the population (Donner and Tellez 2008). There are still many people in Africa who do not have access to traditional banking services. This makes m-banking a practical and highly accessible solution. According to the GSMA, more than half of the 270 m-banking offers marketed worldwide in 2015 were in Africa and more than 84 million active accounts out of 134 million registered worldwide are on the African continent.

## I.2. Governance of digital technologies in Africa

The success of the telecommunications sector in Africa is not without its challenges. Indeed, the increase in Internet use also poses many threats to the economies of several African countries and their image abroad due to the development of certain crimes such as scams carried out using digital tools (cell phones and the Internet) both within countries and at the international level. This situation is the pretext for major campaigns to identify digital technology users in various African countries in order to regulate activities in cyberspace and make it secure. This policy of identifying all mobile telephone and Internet subscribers, initiated by several African countries such as Angola, Gabon, Cameroon, Niger, Kenya and Côte d'Ivoire, appears to be a new situation for all cyberspace users, and in particular for digital public space activists. The opportunity for cyberspace users to express themselves on issues of general interest, enjoying anonymity if necessary, has often been perceived by some citizens as an opportunity to denounce the evils of society in complete serenity and without fear of reprisals.

In short, the popularization of the Internet and its applications in Africa contributes to the diversification of the means of communication in addition to the traditional media. Public use of social networks, for example, raises a number of questions, particularly about the stakes of these uses on an African continent weakened by political violence that is generally amplified by the news media. In these African uses of modern technical and digital communication channels, digital social networks constitute the new challenges of information, debate and mobilization (Breton 2000) of public spaces in reconstruction. These African public spaces, sociability, communication and mediation locations (Breton and Proulx 2006) appear increasingly as a need for citizens to express themselves more freely on the current events that interest them. Social media seems to be an opportunity for African populations, which is a form of freedom of expression that is not without its challenges. In other words, the African public space and the use of social media offer avenues for reflection on the challenges of such social and media interaction links.

In recent years, decision-makers in the UEMOA region have promoted the adoption in African public universities of the Licence–Master–Doctorat (LMD) system (licensure, master's, doctorate), the implementation of which is supported by projects such as the Support Project by Information and Communication Technologies (PADTICE), initiated by the UEMOA and the UNESCO. The integration of ICTs into this innovation is thus identified as an opportunity to assist in capacity building and a viable solution to facilitate the implementation of the reform. However, as Ducasse (2002) pointed out, "the mere connection to a communication system, even the Internet, is not a guarantee of good practice for anyone [...]".

Thus, after several years of operation under the LMD reform, African public universities have revealed various forms of difficulties in appropriating digital tools in the management of services, staff, curricula, teaching and research. It is therefore important to understand the "foundations of the unease" (Gnahoulé-Oupoh 2013) that undermine this reform.

As can be seen, the penetration of digital technologies into African society has many economic, political, social and cultural challenges that the chapters of this book attempt to highlight.

The issues are mainly related to social media and their involvement in democratic construction, or the socio-political development of public spaces on the African continent. Part 1 of this book examines, from an interdisciplinary angle, the modes of occupation of public space and its "deliberative dimensions with regard to socio-political" governance in Africa. Africans, especially the younger generation, have appropriated new means of communication, ICTs, to participate in political debates in their own way by getting involved, informing themselves and denouncing what they consider to be shortcomings in the African democratic game. The development of ICTs in Africa and the diversification of uses therefore allow young Africans to free themselves from the traditional and almost inaccessible media muzzled by public authorities to invite themselves into and make their presence known in the political debate.

The economic aspect of ICT use reveals more major issues than we can address in a single book. The advent of ICTs, such as mobile telephony, since the 1990s, has had a positive impact on African economies, particularly through the development of value-added services such as *m-banking*, distance learning, e-commerce and money transfer. The emergence of these technologies has brought about several changes in Africa, both positive and negative. While several countries have seen the socio-economic conditions of certain segments of the population improving, the phenomenon of cybercrime, which is quite developed in West Africa, constitutes a serious obstacle to this development through the use of the Internet. This deviance in the use of electronic tools is used as a pretext for digital surveillance under the banner of projects to build “smart cities”, protect populations through digital identification and install surveillance cameras in large urban areas.

The concluding part of this book focuses on human development through ICTs, which is a topical subject whose challenges merit scientific curiosity. From higher education (where the quality of training provided has deteriorated due to overcrowding) to hospitals (where the question of population care is increasingly raised), digital technologies are being introduced as stopgaps, which also generate new problems in education and healthcare.

The use of ICTs in Africa constitutes a major development challenge that gives rise to scientific reflections to which all the following chapters submit themselves in an attempt to understand the situation.

## I.3. References

Acker, J. (2008). Does Digital Divide or Provide? The Impact of Cell Phones on Grain Markets in Niger. *Bread Working Paper*, 177(May), 1–60.

Alzoum, G. (2008). Téléphone mobile, Internet et développement: l’Afrique dans la société de l’information? [Online]. *tic&société*, 2(2). Available at: http://ticetsociete.revues.org/488.

Bonjawo, J. (2002). *Internet, une chance pour l’Afrique*. Karthala, Paris.

Breton, P. and Proulx, S. (2002). *L’explosion de la communication. Introduction aux théories et aux pratiques de la communication*. Éditions du Boréal/La Découverte, Montreal/Paris.

Chéneau-Loquay, A. (2008). Rôle joué par l’économie informelle dans l’appropriation des TIC en milieu urbain en Afrique de l’Ouest. *Netcom/Netsuds*, XXI(3–4)/4.

Donner, J. (2006). The Use of Mobile Phones by Microentrepreneurs in Kigali, Rwanda: Changes to Social and Business Networks. *Information Technologies and International Development*, 3(2), 3–19.

Donner, J. and Tellez, C.A. (2008). Mobile banking and economic development: Linking adoption, impact and use. *Asian Journal of Communication*, 18(4), 318–332.

Gnaoulé-Oupoh, B. (2013). Université F.H.B: les fondements du malaise, la découverte de l'arnaque [Online]. Available at: https://www.connectionivoirienne.net/2013/06/01/universite-f-h-b-les-fondements-du-malaise-la-decouverte-de-larnaque/.

Programme des Nations unies pour le développement (PNUD) (2001). Briefing kit [Online]. PNUD-Congo, Brazzaville. Available at: http://ipl.econ.duke.edu/bread/papers/working/177.pdf.

Tall, S.M. (2004). Senegalese Émigrés: New Information and Communication Technologies. *Review of African Political Economy*, 31(99), 31–49.

PART 1

# Social Media and the Public Space in Africa

# 1

# Civil Society in the Consolidation Process of Democracy in Burkina Faso: Balai Citoyen on Facebook

In the early 1990s, some international organizations and non-government organizations (NGOs), including those endorsed by UN bodies and working in development such as CARE, Human Rights Watch, Oxfam, VSF-CICDA, GRET, etc., found a way to evaluate the state of African-style democracy in the concept of *good governance* (Cubillos and Appolin 2004). For these institutions, evaluation measures were based on a number of founding concepts: participation, rule of law, adjustment capacity, consensus orientation, equity, accountability and strategic vision. In this interpretation, *good governance* must make it possible to combine the rights and duties of a community of interests.

In Burkina Faso, civil society organizations (CSOs) have been looking at the management of public affairs by politicians since the beginning of independence in 1960. To this end, we note the following:

> "Beyond the political instability that has always characterized Upper Volta, in the 1970s it was able to afford the 'luxury of democracy', even if it was 'a democracy with a military face' in a regional environment dominated by a single party; it owes it above all to the existence of a combative civil society, which has managed to preserve a minimum level of autonomy in

Chapter written by Marcel BAGARÉ.

> relation to the attempts to control the different regimes that have followed, until the advent of the Burkinabe revolution on August 4, 1983." (Loada 1999, p. 7)

While the political resolutions of the revolution initiated by Thomas Sankara at the end of the coup of October 15, 1987 made it possible to bring civil society closer together, it was above all the democratization process that began in the 1990s that created a favorable context for the expansion of the latter (Sandwidi 1996). From the murder of journalist Norbert Zongo in 1998 to the popular uprising that preceded the fall of the Blaise Compaoré regime in 2014, Burkina Faso is experiencing a crisis of legitimacy and effectiveness, of which abstentionism and mistrust of citizens towards decision-makers are the most obvious symptoms. At the same time, citizens show, more than ever, a desire to get involved in the management of public affairs, and the structures that allow them to invest themselves in the life of the city are increasing in number (Saldomando 2001). According to a European Union report, Burkina Faso has about 12,000 CSOs (professional, religious, youth, trade unions, national and international NGOs, community organizations, networks and umbrella organizations, etc.) (EU+ 2016 Group). Thus, organized civil society, as a lever for revitalizing the democracy, aspires to be fully integrated into democratic decision-making processes (Planche 2004).

The emergence of Burkinabe civil society on the political scene is not new; however, it is with the emergence of *Balai Citoyen* (French for Citizen's Broom) and many others that civil society has emerged as an essential actor in political change. This movement has undergone qualitative development marked by a commitment to the struggle to strengthen the democratic process and build the rule of law. The role played by Balai Citoyen in the period from October 27 to 31, 2014, preceding the attempt made by the Blaise Compaoré regime to amend Article 37[1] of the Constitution, reveals a mechanism for mobilizing all segments of the population in favor of political change (Brother and Angelbert 2015). Thus, through the popular mobilization it provoked, Balai Citoyen undeniably led

---

1 Article 37 stipulates the duration of the term of office and the modalities of the eligibility of the President: "The President shall be elected for five years. He may be re-elected only once" (Constitution of Burkina Faso, 1991). According to the draft law, the text would have been amended as follows: "The President of Faso is elected for five years by direct, equal and secret universal suffrage. He may be re-elected twice".

the regime to include these problems in its political agenda. In this sense, the politicization of issues related to impunity, justice, the rule of law and democracy has led to the integration of Balai Citoyen into the Burkinabe political scene and given its activities a political character.

How can we explain the rapid emergence of Balai Citoyen in the political sphere in Burkina Faso? The choice of social networks (Facebook) on the Internet by Balai Citoyen was a strategic means of mass communication in order to bypass the traditional communication channels under the control of the regime. This general mobilization approach of populations by Balai Citoyen on Facebook is part of a dynamic:

> "to support as closely as possible the evolution of ever more efficient media. Political communication very quickly took hold of the Internet, which has increased its power and considerably extended its area of influence, thanks to its rapid evolution towards Web citizenship, which forms a brand new network of rumors and influence. The relatively recent evolution in the early 2000s from the Internet to Web 2.0 and a much more social and relational set of properties (collaboration, sharing of information and content, personalization, knowledge management, use of social networks, communication) has fostered the emergence of digital social networks that have quickly crystallized politicians' expectations and desires for transparency, immediacy and proximity to citizens." (Babau and Eyries 2015, p. 2)

## 1.1. Problems

Like other social networks, Facebook is a field of projective reality and a construct of the social, political and cultural imagination. In a dynamic of social change, these types of social changes become products accepted as reality by the public. On Facebook, the reality of events is, in most cases, assimilated with the projective reality. Such an approach leads audiences to act by supporting symbolic constructs that have few or almost no link with reality (Palliart 2000). Projective messages, derived from the myths maintained by social networks such as Facebook, encourage its users to collectively live a story with the main actors. Internet users thus contribute to the construction of a virtual reality and convert it into real life. How does Balai Citoyen proceed to build an image of itself as an active player on

Facebook? Through what operational mechanism does Balai Citoyen use Facebook to influence the Burkinabe political landscape after Blaise Compaoré? What role do the key actors of the movement play on social networks in the political struggle and democracy in Burkina Faso? To explore these questions, we will consider two main hypotheses. The first hypothesis concerns the fact that the main actors of Balai Citoyen have been able to conquer the public on Facebook through charming operations that consist of displaying their socio-professional status. As for the second hypothesis, we note that Balai Citoyen uses Facebook as a tool for political combat in defending the democratic ideal through crowd mobilization.

## 1.2. Theoretical and methodological approach

The objective of this study is to show the role of Balai Citoyen in the consolidation of post-insurgency democracy in Burkina Faso through social networks (Facebook). It is part of a systemic approach that builds on the work of the invisible college of the Palo Alto school. The systems theory developed by this school reflects the action of one element on another, which in turn leads to a response from the second element to the first. This is what we call feedback, as highlighted in cybernetics that lays the foundation for this theory. This approach not only studies a few variables in isolation, but also reflects an overall interaction situation. In this reflection, social actors (here, Balai Citoyen) move themselves away from McLuhan's determinism and are granted a new role. It is acknowledged that technology (here, Facebook) is in turn influenced by its users. From this circularity of the communication scheme arise uses that, to a certain extent, become the identity of technologies. Initially receptors, social actors become influencers by developing tricks to adapt to their new environment. From the interactive communication that is established between the two entities, there is a phenomenon of readjustment of the Balai Citoyen actors to Facebook technology and of the latter to its uses. Thus, to answer the questions raised by this reflection, we chose to conduct a mixed study (qualitative and quantitative) by analyzing the content published by the main Balai Citoyen actors during the post-insurgency period, from November 17, 2014, the date of appointment of the President of the transition, until after the presidential elections of November 29, 2015 on their Facebook account. The duration of the research is therefore 12 months, or one year (from November 17, 2014 to November 29, 2015), and corresponds to the post-insurgency period, which includes a major event: the presidential election campaign, which symbolizes the first steps towards democracy. This period was chosen because the messages became more crystallized and it became easier to analyze them.

We observed from the beginning to the end the Facebook accounts of three main leaders of Balai Citoyen and the accounts of users who claimed to be part of this movement. We used qualitative content analysis, or anchored theorization analysis, to study their messages (Guillemette and Luckerhof 2009). Qualitative content analysis is a method that allows the inductive generation of a theory on a cultural, social or political phenomenon through the conceptualization and progressive linking of empirical data. The particularity of this method is its independence: qualitative content analysis can be used as a general research strategy because it maintains a certain degree of theoretical autonomy. Indeed, in this type of analysis, it is not necessary to refer to or respect the principles of a known theory to interpret the data. Whatever the dimension of the analysis, what we obtain will always be anchored in the empirical data. The empirical material is at the same time the starting point for theorizing, the place where emerging hypotheses are tested, and the last test that must validate the empirical construction as a whole.

Our quantitative approach is based on the production of quantitative data on the observed phenomenon, namely the use of Facebook by Balai Citoyen as a means of political struggle. This makes it possible to know the proportions of a given phenomenon and thus to have a clear idea of the representativeness of opinions, practices, etc. or the correlation between two given phenomena (Balima and Veronica 2005). For our study, we will carry out a categorical analysis of the constitutive meaning modalities of Facebook publications, commonly known as posts. Category analysis "consists of calculating and comparing the frequencies of certain characteristics (most often the themes mentioned) previously grouped into significant categories" (Quivy, Compenoudht and Marquet 2017, p. 232). It is based on the frequency of occurrence of certain content characteristics or their correlations. For the operationalization of this method, we use the observation technique.

The main method used to determine the corpus of posts to be analyzed is the census of all the content published by all the Balai Citoyen actors on their accounts during the period of our study, i.e. the insurrectional posts also including the presidential election. We have removed all personal posts that do not contribute to an understanding of the phenomenon under study. In this phase of corpus development, we identified two categories of actors.

The first category concerns the three charismatic leaders of the movement. The second category is mainly made up of representatives and supporters of Balai Citoyen, scattered throughout Burkina Faso. We have identified nine of them. The corpus also concerned posts that we categorized according to the nature of the published content. Three types of post content emerge from this step: photos, videos and comments. Table 1.1 gives an overview of the general corpus of the study.

| **Categories of actors** | **Sample** | **Post content** | | | **Total** | **%** |
|---|---|---|---|---|---|---|
| | | **Videos** | **Photos** | **Comments** | | |
| **Leaders** | 3 = 21.43% | 38 | 62 | 136 | 236 | 51.52 |
| **Representatives/ supporters** | 11 = 78.57% | 54 | 76 | 92 | 222 | 48.47 |
| **Total** | 14 = 100% | 92 | 138 | 228 | 458 | 100 |
| | | 20.09% | 30.13% | 49.78% | | |
| | | 100% | | | | |

**Table 1.1.** *Distribution of post content by actor (source: field survey conducted by the authors in 2018)*

## 1.3. Good governance, the purpose of Balai Citoyen

Since the 2010s, Africa has been generally experiencing a wave of political unrest against the backdrop of constitutional reforms that are ill-suited to the democratic context. This environment has encouraged the emergence of a new type of citizen movement across the continent, which is increasingly focused on the management of public affairs by the authorities in power (Frintz 2014). Among other things, we have witnessed the birth of movements such as "*Y'en a marre*" (I'm sick of it) in Senegal, "*Trop c'est trop*" (Enough is enough) and "*Ça suffit*" (That's enough) in Chad, "*Filimbi*" and "*Lucha*" in the Democratic Republic of Congo, etc. In Burkina Faso, the Balai Citoyen movement was created in June 2013 in pre-election political observations of 2015. Its first actions were carried out jointly with the political opposition through protest marches against the constitutional reform aimed at setting up the Senate. The Balai Citoyen movement can be identified through many slogans that express the break with a political system considered tribal. We can read on signs, during demonstrations, slogans such as: "*notre nombre est notre force!*" (Our size is our strength!)

and "*ensemble, la rupture n'est jamais seule*!" (Together, a breakup is never lonely!). This movement is also characterized by its local and decentralized organization and structure. At the head of the movement, we note the presence of two committed musical artists: the rapper Smockey (real name Serge Martin Bambara) and the reggae artist and radio host Sams'K Le Jah (real name Abdoul Karim Sama), a native of Côte d'Ivoire, nicknamed "the man of the ghetto". In 2018, they recorded an album together called *Norbert Zongo: dossier classé*, to denounce the abuse of the Blaise Compaoré regime after the death of the brutally murdered journalist (Gorovai 2016, p. 523). According to the author, in addition to these two artists was the lawyer Guy Hervé Kam, President of the Centre for Judicial Ethics. He was one of the initiators of a petition launched in 2010 calling for the inviolability of the constitutional article (Article 37), which set the number of presidential terms at two. It was he who became the spokesperson for Balai Citoyen and drafted its charter. Apart from these main actors, Balai Citoyen has a pyramidal structure. On the territorial level, there are clubs of citizens composed of associate members. At the higher level, there are focal points that are subordinate to regional coordination. Finally, at the national level, there is the annual National Assembly, which designates the national coordination, composed of 13 members, including graduate students, shopkeepers, three musicians, two journalists and a lawyer. Balai Citoyen has 60 clubs in the capital Ouagadougou and about 40 others, deployed in different parts of the country.

The main objective of Balai Citoyen is to fight against the abuse of the Blaise Compaoré regime, after its attempt to subvert the constitutional framework to be able to run for President 2015. For the initiators of the movement, this struggle aims to work towards popular resistance and a better future for the country through social and political change that respects the rules of the republic and democracy[2]. Compaoré's Burkina Faso was a typical example of the hegemonic regime, the semblance of a democratic game aimed only at locking down the system. What will become the Congress for Democracy and Progress (CDP) was formed in the aftermath of Sankara's assassination by Compaoré's men in 1987 as a political–military apparatus. The economy was then dominated by the networks of the ruling

2 Élections 2015: Le Balai Citoyen lance sa campagne "Je vote et je reste" [Online]. November 6, 2015. Available at: http://lefaso.net/spip.php?article67830 [Accessed July 22, 2017].

clan[3]. Le Balai Citoyen was born from the observation that nothing positive came out of this partisan game. All CSOs, including the Balai Citoyen and political parties, worked towards the fall of the regime in 2014, including the *Union pour le progrès et le changement* (UPC) of Zéphirin Diabré, the *Union pour la renaissance-Parti sankariste* (Unir PS) of Bénéwendé Sankara and above all the *Mouvement du peuple pour le progrès* (MPP) of Roch Marc Christian Kaboré, Salif Diallo and Simon Compaoré, three leaders of the CDP who stepped down in January 2014 to create the MPP (International Crisis Group 2013). In the period leading up to the departure of Blaise Compaoré's regime, the Balai Citoyen wished to maintain its status as a civil society organization and formed, with other similar structures, the *Coalition des organisations de la société civile* (COSC) (Soré 2016).

To reach its citizens, Balai Citoyen uses both traditional media and information and communication technologies (ICTs). It organizes press conferences, makes statements in the written press and conducts interviews on radio and television. However, it is most talkative on social networks. The communication unit regularly produces publications on Facebook and Twitter. These accounts make it possible to quickly inform activists and sympathizers about the movement's current events, particularly its positions on national political issues.

## 1.4. The media construction of Balai Citoyen via Facebook: images and propaganda

In a political context free of any form of authoritarianism, the transition following Blaise Compaoré's departure was a zone of induced turbulence by political unrest for the conquest of power. If there is one sector that has escaped political upheaval, it is the media sector that has enjoyed some freedom of expression since the death of journalist Norbert Zongo. It is therefore an alternative public space that allows information to flow nationally and internationally (Yabara 2015). It has also been able to facilitate the coordination of protest rallies. However, in the context of our study, social networks (Facebook), like traditional media, have helped to spread a very simplified vision of a necessarily complex post-insurgency

---

3 BOA, Burkina Faso: "Le Balai citoyen", un mouvement pour 'assainir' la mal-gouvernance [Online]. July 18, 2013. Available at: http://koaci.com/burkina-faso-balai-citoyen-mouvement-pour-assainir-gouvernance-83932.html [Accessed September 2, 2017].

political process, often focusing on an explanatory variable and ignoring the different sequences of this process.

| Audiovisual content | | | | | | |
|---|---|---|---|---|---|---|
| **Actors** | **Visuals/photos** | | | **Audiovisuals** | | |
| | **Individual** | **Collective** | **Events** | **Individual** | **Collective** | **Events** |
| **Leaders** | 27 = 43.54% | 21 = 33.87% | 14 = 22.58% | 9 = 23.68% | 18 = 47.36% | 11 = 28.94% |
| **Total** | 62 = 26.95% | | | 38 = 16.52% | | |
| **Representatives/ supporters** | 18 = 33.33% | 25 = 46.29% | 11 = 20.37% | 11 = 14.47% | 43 = 56.57% | 21 = 27.63% |
| **Total** | 54 = 23.47% | | | 76 = 30.04% | | |

**Table 1.2.** *Individual and collective actors involved in post content*

After observing Balai Citoyen posts on Facebook, we recorded results confirming that the Balai Citoyen movement has built its reputation on the brand image of leaders enjoying great popularity, due to their professions and socio-political commitments. To this end, the results in Table 1.2 show that the image of the three leaders who embody the Balai Citoyen movement constitutes an important part of all audiovisual posts published on Facebook. The three leaders, who represent 21.43% of all active actors (14) of the Balai Citoyen movement on Facebook, used their image in audiovisual publications in the following proportions: 26.95% photos and 16.52% videos, representing a total of 43.47% of the audiovisual posts of all active actors of Balai Citoyen. This disparity noted on Facebook shows that the Balai Citoyen movement uses the so-called e-reputation strategy, or reputation on the Internet, to win over sympathizers. Concept-reputation is defined as:

> "a notion of sensitivity to the information provided, of predisposition in relation to the message designed by the brand. It is this sensitivity, the socio-cultural frame of reference and the perception of each individual that will make it possible to form an opinion on a brand. The target audience will react to a message, taking into account, consciously or unconsciously, these different elements and form an opinion. It is this opinion that will contribute to the development of the e-reputation of

> a brand: it is the sum of the opinions of the whole with regard to the brand." (Tamseddak 2107, p. 35)

While this concept is generally used in the field of marketing to popularize the brand image of products, it has been transposed into the political sphere as a propaganda tool for the image of actors.

In the context of this research, e-reputation consists of the opinion and feeling towards the image of the Balai Citoyen actors on Facebook. It is defined here by visibility (number of photos or videos) and qualification (number of "Likes" for photos and number of "views" for videos). However, e-reputation not only depends on online content but is also subject to offline or real life actions. Thus, the various actions undertaken by the Balai Citoyen actors, whether they are campaigns, events or partnerships, have an impact on e-reputation. The slightest word spoken and relayed can influence opinion, for good or for bad. The border between online and offline is porous and one influences the other constantly. Indeed, everything can be taken online and influence an image that has been worked on beforehand. In our study, actors construct a digital identity for specific purposes. Digital identity can be defined as "all the contributions and traces that a company (or person) leaves online, voluntarily or not" (Benghozi, Bergadaam and Burkhart 2011). Virtual identity, within the framework of this work, is based on two main types of data that can be observed on Facebook: formal data from official bodies such as institutional sites (government, companies, etc.) and informal data that generate voluntary contributions (publication on social networks, forums, etc.).

Regarding content analysis, the publication of texts by the Balai Citoyen actors on Facebook shows their involvement in the political arena in Burkina Faso. This study focuses on the nature of these actors via Facebook and how they manage to appropriate this social network in order to express their civic commitment to democracy. In our context, it is a flow of publication posts and a flow of comment posts, which causes interactivity on the pages of the Balai Citoyen actors. While the maxim of the Palo Alto school confirms that we cannot fail to communicate, the Balai Citoyen actors, for their part, have communicated widely during the post-insurgency period. Citizens took the opportunity to express their opinions freely. Through content analysis, it appears that Facebook has participated in the construction of a new democratic era in Burkina Faso. It is about reconfiguring the social bond, or even rebuilding the relationship between politicians and citizens.

There is no longer any room for authoritarian discourse and we are talking about a place of interactive public citizen debate, where the politician expresses his or her ideas and the citizen comments. Citizens represent a virtual community that pursues the same objective: to express themselves freely *through* the virtual space that is Facebook. The analysis presented in Table 1.3 shows the diversity of actions and themes of the discourses of the Balai Citoyen actors. Balai Citoyen is not a political structure, but carries political claims. In this way, it makes citizen mobilization its main focus. The movement's actions mainly fall into the following three main categories: mobile citizen information tours, educational concerts and sit-ins and march meetings. The main actions of the movement after October 2014 will be highlighted from three specific moments that mark the transition:

– during the post-Compaoré negotiations;

– during the failed coup in 2015;

– in the electoral context in 2015.

With regard to the aspects criticized and the demands, we find the following general themes: social injustice, justice, alternation, good governance, protection of citizens' rights and corruption.

| **Themes** | **Actors** | | **Total** | **%** |
|---|---|---|---|---|
| | **Leaders** | **Representatives/sympathizers** | | |
| **Social injustice** | 17 | 4 | 21 | 9.21 |
| **Justice/ impunity** | 40 | 26 | 66 | 28.94 |
| **Change in power** | 26 | 25 | 51 | 22.36 |
| **Good governance/ mismanagement** | 14 | 11 | 25 | 10.96 |
| **Corruption** | 19 | 20 | 39 | 17.1 |
| **Protection of civic law** | 20 | 6 | 26 | 11.4 |
| **Total** | 136 | 92 | 228 | 100 |
| | 228 | | | |

**Table 1.3.** *Distribution of themes according to the actors on Facebook (source: field survey conducted by the authors in 2018)*

## 1.5. Leadership: a factor of mobilization via social networks

In the search for an applied theory for political communication in a format that touches the mass media, Agenda Setting can be entrusted with this mission with the emergence of social networks. This theoretical approach applies to the media's ability to select, prioritize and generate standards for information development (Ganjon 2010).

Thus, the theory attempts to explain the crucial role of the media in the dissemination, selection and concealment of information, and also in the importance that the public in turn attaches to these media. This trend gives power back to the media, after the predominant idea that their effects were limited. This theory is based on the assumption that the media give more importance to certain facts than others. They systematically try to control the public's choice of themes that they have prioritized and that will shape public opinion (Gracias and Naddaf 2016).

One of the themes developed by Balai Citoyen on Facebook is justice and impunity for crimes committed since the end of the 1987 popular revolution. In this dynamic, the Balai Citoyen movement promotes the image of the leader of this revolution, Captain Thomas Sankara, "elevated to the rank of supreme cibal" by virtue of the fact that it constitutes the "national political heritage, source of inspiration for the cibals"[4]. The use of the portrait (photo) of the leader of the Burkinabe revolution Thomas Sankara on communication media (posters, tarpaulins, banners, t-shirts, etc.) during demonstrations (mobile citizen information tours, conferences, sit-in and march-meetings) is widely visible on Facebook pages of both the movement's leaders and other representatives. Table 1.3 reports that on the theme related to the concept of justice, the content of speeches (texts posted) on the actors' Facebook pages represent 28.98% of the entire corpus. This use of Thomas Sankara's effigy on social networks by the Balai Citoyen leaders is of a strategic nature. The post-insurgency period was a period of reshaping the political landscape among all actors, including those of Blaise Compaoré's party (CDP). For the Balai Citoyen movement, all those who supported Blaise Compaoré until the last few hours in his endeavors to remain in power against the will of the people are also guilty of the crimes committed, including the murder of Thomas Sankara. The very marked

4 Guide ou manuel du Cibal, publié sur Facebook [Online]. Available at: www.facebook.com/notes/1446787425547753/ [Accessed June 24, 2017].

presence of the latter's effigy on Facebook in the publications of the movement's leaders and other representatives was intended to influence citizens in the choice of political actors during the upcoming elections. Moreover, in the Balai Citoyen vocabulary, reference to honest men occurs regularly. In the elements of the corpus, the posts (texts, images and videos) and arguments focus on the figure of Sankara, which is associated with Marxist–Leninist ideology, according to populist and pan-Africanist principles.

Thomas Sankara's image is therefore a symbol that carries values on several levels. The first aspect that emerges in our study is the interest of Burkinabe youth in the leader of the revolution through his commitment and his ideology. For the Balai Citoyen movement, he is considered a "Che Guevara" figure, the emblematic leader of the Cuban revolution, who was also assassinated. Thomas Sankara becomes for the youth, especially for the Balai Citoyen leaders, the embodiment of the revolutionary who fights, for the people, against Western imperialism. Moreover, on all the audiovisual elements of the corpus of our study, and on all the actors of the Balai Citoyen movement, we find a black t-shirt bearing the effigy of Che Guevara against a background of the colors of the Burkina Faso flag, referred to Thomas Sankara. It is therefore a revolutionary leader who hides another. Then, the movement's own approach finds justification in the system promoted by Sankara.

Another reason for the use of the theme of justice by the actors of the Balai Citoyen movement on Facebook is the disappearance of another leader: the journalist Norbert Zongo, who was also murdered by the Blaise Compaoré regime without justice being served. On the anniversary of his death, on December 13, celebrated each year, the Balai Citoyen movement relays information about its remembrance activities on Facebook. The communication media used by Balai Citoyen, including Facebook during this period, include the slogan: "Justice for Norbert Zongo". Given the interest of young people in the Norbert Zongo case, this is another opportunity to put pressure on all the employees of the former Compaoré regime who work in the political sphere. The results of the body observation report that all leaders and representatives of the Balai Citoyen movement wore black T-shirts in the movement's colors with the inscription "Justice for Norbert Zongo" during these commemorations. Another case highlighted on social networks via Facebook by Balai Citoyen is that of a leader from the Burkinabe students' union (ANEB), named Dabo Boukary, whose disappearance in previously unknown conditions is attributable to the same Compaoré regime. Balai Citoyen participated in the commemoration of the anniversary of his disappearance to show their support for the student community in the search for truth and justice. The texts, photos, videos and comments of events posted on Facebook by the movement contribute, in one way or another, to influencing the landscape and political debate in a context of post-insurgency electoral competition.

For political scientist Michel Hermans, Facebook is used by the political world as a unilateral propaganda tool:

> "What social networks have changed in the political world is the constant possibility of showing pictures of oneself, to publish articles, interpellations, show what they have done, be nice or share their stories; without having to go through traditional media." (Hermans 2005, p. 24)

A total and complete control of one's image remains illusory, if not impossible, since it is built not only according to one's own actions and statements, but also according to interactions with members of social networks. One of the strategies for mobilizing Balai Citoyen is the use of popular figures like Thomas Sankara, essentially on moral positions. It also draws inspiration from other African leaders (Nelson Mandela, Nkrumah, Lumumba, Cheikh Anta Diop, etc.), pacifists (Gandhi, Martin Luther King) and committed artists (Bob Marley, Alpha Blondy, Tiken Jah Fakoly, etc.). This form of use is called marketing use (Verville and Giasson 2011). Although this concept has been widely studied in the field of political advertising and media coverage (Kriesi 2012), it is now a form of use adopted by political actors on digital platforms, in this case Facebook. In the same context, G. S. Enli and Skogerbø (2013) believe that social media has entered into a process of political communication based on the personalities and characteristics of political actors. This is similar to the concept of political mediatization by Esser and Stromback (2014), who argue that political actors must build, share a public image and value their personal and private lives. It is in this same dynamic that Balai Citoyen is putting itself forward in its Facebook activities. Its leaders (Sams'K Le Jah and Smockey) are very popular with young people (see Table 1.2). This movement is perceived by its designers, Sams'K Le Jah and Smockey, as "a vehicle for mobilization and resistance in a political context that creates and instigates the exclusion and marginalization of the greatest number of Burkinabe citizens to the benefit of a minority that grants itself undue privileges"[5]. The artistic talent of the two leaders, who excel respectively in reggae and rap with committed themes, is an asset for mobilization during the educational concerts organized by Balai Citoyen. Indeed, Facebook induces a participatory culture that increases the loyalty links between the Balai Citoyen actors and the populations and provokes a communication focused on interactivity, especially since there is a democratic unease and a decrease in political interest and participation in Burkina Faso (Chawick and Stromer-Galley 2016).

## 1.6. Facebook and Balai Citoyen: an instrument of conquest and political struggle?

The role of the Balai Citoyen movement is to contribute to liberalizing and promoting public discourse of general interest in a political context crystallized by the consequences of the Compaoré regime. It is part of the dynamics of CSOs and appears as a torchbearer of democracy by distinguishing itself from any political role. In this last part, the Balai Citoyen publications on Facebook during the study period reveal three types of actions based on the events that marked this phase of Burkina Faso's history. After October 2014, the date of Blaise Compaoré's departure from power, Balai Citoyen was effectively illustrated through three types of acts:

– awareness-raising actions during negotiations for reshaping the political landscape;

– acts of protest and denunciation during the 2015 coup orchestrated by the RSP under the transition;

– acts of appeal for transparency and fair play in the political game in the electoral context of 2015.

The first posts of Balai Citoyen on Facebook during this stage aimed to justify its role as an impartial actor in the political game after the military took power with Colonel Isaac Zida. To this end, several posts on social networks (Facebook) expressed suspicions of connivance between the Balai Citoyen movement and the military. At the root of this controversy were images and videos posted on Facebook, where some of the leaders of Balai

5 Available at: www.koaci.com/burkina-faso-balai-citoyen-mouvement-pour-assainir-gouvernance-83932.html [Accessed November 9, 2017].

Citoyen were displayed alongside the one who claimed to be the leader of the transition during his first public appearances. The multiple reactions of accusation of Internet users on Facebook led the leaders of Balai Citoyen to hold awareness conferences on their role as mediators between political actors, the military and the public. The first publications of the movement on Facebook to reassure Internet users about their non-involvement in the political game rejected the appointment of a transition minister who was suspected of having served time in prison in the United States, following embezzlement. The various posts covered almost all of the movement's actions in the field. Balai Citoyen recorded a victory with the resignation of the latter, following numerous demonstrations held in front of the ministries. Balai Citoyen continues its actions and focuses on its role as a sentinel, ensuring that the transition runs smoothly. The movement is also involved in citizen dialogue at the local level, which is based on debates around political agendas. This type of positioning indicates that Balai Citoyen begins to take its responsibilities according to its objectives and to progressively establish itself as an actor of common interest (Emiliano and Sabine 2006).

The second event is related to September 17, 2015, when representatives of the RSP[6] announced the overthrow of the transitional government, led by President Michel Kafando and Prime Minister Isaac Zida. As soon as this announcement was made, Balai Citoyen reacted on Facebook by calling on Internet users to mobilize, in order to block the way to this adventure of soldiers from the ranks of the RSP. Balai Citoyen gave strict instructions on Facebook. These instructions were undoubtedly similar to a strategy of unarmed struggle to prevent a coup. On the Facebook pages of its members, Balai Citoyen called for the general mobilization of all "comrades", with caution and calmness, and even to hide their sympathy for Balai Citoyen by masking all the distinctive signs that could identify them. The movement called for civil disobedience and the use of all possible means to defeat the putschists they called "terrorists".

Several messages were launched by Balai Citoyen on Facebook, which focused on several areas:

– the "general alert", which referred to the ongoing military coup and called for people to "mobilize and fight!";

---

6 The *Régiment de sécurité présidentielle* was a unit of the armed forces of Burkina Faso, which was responsible for the protection of the President, republican institutions and any other person designated by the President of Faso from 1995 to 2015.

– the “call for popular resistance”, which addressed the people of Burkina Faso to inform them of the RSP’s attempts to “halt the transition and restore the old regime”; call for the immediate release of the President, Prime Minister and ministers;

– the call on the defense and security forces (FDS) to respect their sovereign duty, as well as the population to organize themselves in their own streets, neighborhoods and sectors in order to prevent the RSP from achieving their objectives.

While, in the end, such a strategy bore fruit with the failure of the coup, it did not lower the guard. On November 19, alerts were sent on Facebook by the Balai Citoyen movement against demobilization because RSP putschists could leave the country to escape justice. This Balai Citoyen strategy was part of its “citizen watch” dynamic, which did not exclude the acts of mobilization of a revolutionary nature. The failure of the coup led to a new planning of electoral deadlines, which constituted a favorable period for all political manipulations, which once again mobilized Balai Citoyen in its role as a scout.

In this electoral process, Balai Citoyen was committed to ensuring that the choice of new authorities was made in strict compliance with democratic values, i.e. the transparency of the ballot boxes. The principle of citizen monitoring could not, for its members, weaken; on the contrary, it was critical to maintain the vigilance of citizen control over the transition authorities. It was therefore in this context that Balai Citoyen launched several campaigns on Facebook concerning the vote. The first campaigns consisted of raising awareness among young people about the interest and issues of the electoral process. A slogan was designed for this purpose: “After the revolt, your vote”. This slogan was displayed on all posts of Balai Citoyen’s main actors and was relayed on social networks (Facebook) by supporters and other interested people. The movement understood voting as a civic duty. Subsequently, Balai Citoyen continued its approach through other complementary actions: registration on the electoral rolls and participation in the elections. Balai Citoyen’s posts on social networks, with the slogan “I vote and I stay”, drew attention. This movement campaign on Facebook aimed to achieve a high turnout on polling day and organize an observation of voting and counting, a public process in Burkina Faso. This strategy was accompanied by a concert-meeting with the meaningful participation of the Secretary-General of the Independent Electoral Commission and the transitional government authorities. According to the corpus analysis, overall, there was a significant participation of Balai Citoyen in the electoral process. Indeed, it recommended sending nearly 500 representatives according to the *Convention des organisations de la société civile pour l'observation domestique des élections* (CODEL). During this period, several events were recorded on the Facebook account of the Balai Citoyen actors. On November 9, 2015, Guy Hervé Kam and Smockey called on candidates to avoid ethnic and religious references in the campaign. After this call, Balai Citoyen conducted citizen debates on candidates' programs to raise voter awareness in an educational way. At the end of this free and uncontested electoral phase, the following results were achieved: in the first round, 53.49% of the votes were cast for Roch Kaboré, candidate of the MPP.

## 1.7. Conclusion

Referring to social media as a medium of communication, Facebook offers many possibilities, both for political groups and for the public and the democratic system in general. Connectivity and interactivity are the foundations of this support, in the sense that it facilitates the online participation of users (the user is himself or herself an actor and co-producer of information). In connection with this dynamic, Facebook allows its users to address one or more recipients, to question them on a given issue and to involve them in a specific area. This new communication system will transform the public space of participation and interaction between political elites and citizens (Vergeer and Hermans 2013). At the end of this reflection and in the light of the results recorded, we can consider that the hypotheses developed previously have all been verified: Facebook appears as a social space that favors the construction of a community of common interests based on so-called friendship relationships that are not so in reality. The majority of users of social networks (Facebook) who distinguish themselves as such (friends) have become volunteer emissaries, as is the case in our study of Internet users (users) for Balai Citoyen, which is considered as a brand or even a model. The use of the image of several African political leaders and revolutionaries on the communication media of Balai Citoyen reflects a strategy of struggle that has its roots in e-reputation. In this sense, Facebook has been used by Balai Citoyen for its promotion and control of its image. This use is referred to as "marketing use" by Verville and Giasson (2011). Facebook users have fought for and with Balai Citoyen on the political front, as if this civil society organization were a member of their family.

Beyond the popular uprising, the Balai Citoyen movement was an interesting case to analyze on social networks (Facebook), because it overcame the crisis period of the Compaoré regime by transforming to become a pressure actor with regard to its role during the transition (reshaping of the political landscape and electoral context in 2015). As we have seen, the growth and use of social networks is exponential. This trend is likely to accelerate further in the coming years, with the improvement of the technical capabilities of the Internet and the increased speed of information transmission via broadband. In this case, practices could be modified in a way that is not yet possible to identify.

## 1.8. References

Babau, H.M. and Eyries, D. (2015). Communication politique et réseaux sociaux: les éléction présidentielles 2014 en Roumanie sous la loupe de Twitter et Facebook. *Dossier: Communication – Technologie et développement*, 2(October), 97–111.

Balima, S.T. and Duchenne, V. (2005). *Méthodologie de la recherche en sciences de l'information et de la communication, l'élaboration du mémoire de maîtrise.* Sankofa/Sidwaya, Ouagadougou.

Chawick, A. and Stromer-Galley, J. (2016). *Digital media, power, and democracy in parties and election campaigns: Party decline or party renewal?*. SAGE Publications, Los Angeles.

Cubillos, C. and Appolin, F. (2004). Renforcement de la société civile, politique d'alliances et partenariat. *Traverses*, 14.

Emiliano, G. and Sabine, S. (2006). *Les Groupes d'intérêt. Action collective et stratégies de représentation.* Armand Colin, Paris.

Enli, G.S. and Skogerbø, E. (2013). Personalized campaigns in party-centred politics: Twitter and Facebook as arenas for political communication. *Information, Communication & Society*, 757–774.

Esser, F. and Strömbäck, J. (eds) (2014). A Paradigm in the Making: Lessons for the Future of Mediatization Research. In *Mediatization of Politics: Understanding the Transformation of Western Democracies.* Palgrave Macmillan, Basingstoke, 3–28.

Frère, M.-S. and Angelbert, P. (2015). Briefing Burkina Faso: The fall of Blaise Compaore. *African Affairs*, 114(455), 295–307.

Frintz, A. (2014). La jeunesse africaine bouscule la Francafrique. *Le Monde Diplomatique*, December, 8–9.

Ganjon, F. (2010). Le non-usage de l'Internet: reconnaissance, mépris et idéologie. *Questions de communication*, 18, 37–61.

Gorovai, D. (2016). Le rôle des mouvements citoyens dans le processus éléctoral en Afrique subsaharienne: le cas du Balai citoyen. *SSOAR, Open Access Repository*, 4, 511–537.

Gracias Naddaf, F. (2016). Réseaux sociaux d'Internet: pratique électorale et usage politique de la classe politique chilienne. PhD thesis, University of Paris VII Vincennes-Saint-Denis.

Groupe UE+ (2016). Burkina Faso, Engagement de l'Union européenne envers la société civile: feuille de route 2017-2020 [Online]. Available at: https://www.google.com/url?sa=t&rct=j&q=&esrc=s&source=web&cd=1&ved=2ahUKEwigioa2lL_lAhVxlFwKHfSEBPkQFjAAegQIABAC&url=https%3A%2F%2Feuropa.eu%2Fcapacity4dev%2Ffile%2F83913%2Fdownload%3Ftoken%3DjHd9BbDC&usg=AOvVaw3Ta3O-Q3JH3TGamuOSGems [Accessed 26 October 2019].

Guillemette, F. and Luckerhof, J. (2009). L'introduction à la méthodologie de la théorisation enracinée (MTE). *Recherche qualitative*, 2(28), 4–21.

Hermans, M. (2005). *Internet: le nouvel outil du pouvoir transnational*. Presses universitaires de Liège, Liège.

International Crisis Group (2013). Burkina Faso: avec ou sans Blaise Compaoré, le temps des incertitudes. Rapport Afrique, July.

Kriesi, H. (2012). Personalization of national election campaigns. *Party Politics*, 825–844.

Loada, A. (1999). La réflexion sur la société civile en Afrique: le Burkina Faso de l'après-Zongo. *Politique africaine*, 76(April), 136–151.

Palliart, I. (2000). Les enjeux locaux de la démocratie électronique. *Hermès, La Revue*, 26/27, 29.

Planche, J. (2004). Accompagner l'émergence et le renforcement des sociétes civiles. *Coopérer aujourd'hui*, 38.

Quivy, R., Compenoudht, L.V., and Marquet, J. (2017). *Manuel de recherche en sciences sociales*. Dunod, Paris.

Saldomando, A. (2001). Quelques interrogations sur la gouvernance. In *Les non-dits de la bonne gouvernance*, Haut commissariat à la coopération internationale (ed.). Karthala, Paris, 95–117.

Sandwidi, K. (1996). Syndicalisme et pouvoir politique: de la répression à la renaissance. In *Le Burkina entre révolution et démocratie*, Otayek, R. *et al.* (eds). Karthala, Paris, 325–331.

Soré, Z. (2016). Le Balai citoyen. *Analyse sociétale africaine,* December.

Tamseddak, R. (2007). Le rôle des réseaux sociaux dans l'amélioration de l'e-réputation de l'entreprise. Étude de Cas: DJEZZY. Master's thesis, EHEC, Alger.

Vergeer, M. and Hermans, L. (2013). Campaigning on twitter: Microblogging and online social networking as campaign toolsin the 2010 general elections in the netherlands. *Journal of Computer-Mediated Communication*, 399–419.

Verville, M. and Giasson, T. (2011). Les partis politiques québécois à l'heure du Web 2.0 et de médias sociaux. *European Journal of Communication*, 310–327.

Yabara, A.B. (2015). *Africa Unité, histoire du panafricanisme.* La Découverte, Paris.

# 2

# Social Networks, New Barometers of Political Action in Côte d'Ivoire

## 2.1. Positioning of the problem

Social networks have been growing rapidly for several years all over the world and, of course, in the Global South. Today, it is unthinkable to envisage a society, whatever it may be, without the practice and use of ICT.

In Côte d'Ivoire, according to the ARTCI (*Autorité de régulation des télé communications en Côte d'Ivoire*), it is estimated that the total number of Internet subscribers in the first quarter of 2015 in Côte d'Ivoire was 7,222,261, of which 7,155,583 were 3G mobile Internet subscribers, representing a proportion of 99.089%. According to ARTCI data, the overall Internet penetration rate, which is 31.65%, is boosted by mobile Internet to 31.36%[1].

While in its early days, the Internet was mainly used to exchange information, its uses have evolved considerably and it would not be an exaggeration to talk about an explosion of uses. Social networks have become spaces for meetings, exchanges, expression, mobilization and advocacy.

A few decades ago, in many countries, freedom of the press was a problem.

---

Chapter written by Raymond Kouassi KRA.

1 ARTCI website, available at: www.artci.ci [Accessed August 5, 2016].

With the development of ICTs and the diversification of uses, what was impossible or difficult to express in traditional media (print, radio or television) is now easily said. There are many examples, both in countries with democratic regimes and in kingdoms and other democratizing[2] states.

In this chapter, we will investigate the role of social networks in the political arena in Côte d'Ivoire based on three real-life situations:

– The first case concerns a decision by the Ivorian government, through its Ministry of Transport, to change driving licenses; it should be recalled that the licenses in question were issued less than 10 years ago. They replaced the old three-part licenses. When the decision was made to change them more than a decade ago, the government argued that these new permits were more reliable and more secure because they were biometric. The taxpayer was therefore surprised to hear the same government decide to change them, because they are now "unreliable and not linked to a database"[3].

– The second case relates to what has become known as the "Awa Fadiga case". She is a young woman, a model, who was assaulted in a transport vehicle and died in the emergency room of the Cocody University Hospital (a commune in Abidjan's capital). The death of this young lady caused outrage and consternation both in the traditional media and on social networks.

– The third case is related to the increase in electricity costs in Côte d'Ivoire. This increase, authorized by the government and not announced to consumers, has led to violent street demonstrations in many Ivorian communities[4]. The government had to call for calm and deploy significant security measures to restore order.

In the three situations just mentioned, the government had to back down and temporarily suspend some measures. When we look at these cases through the prism of the Net, we observe that there was a strong mobilization of Ivorians on social networks.

2 We must understand here that developing countries have not yet liberalized their audiovisual space. States still retain control over the flow of information through their media, the state media, known as the public service media.

3 Remarks by the Ivorian Minister of Transport.

4 Serious demonstrations took place in the city of Bouaké, the second largest city in Côte d'Ivoire, leading to material damage, theft and looting of public and private buildings.

It would be presumptuous to blame social networks for the government's retraction on these three issues. But it cannot be denied that the amount of information produced on social networks about these problems has played an important role.

We propose to answer the question of the role played by digital social networks. Are they instruments for measuring social tensions that can become more widespread?

This main question begets three more specific questions:

– What are the contents of the messages that have been posted on social networks about these cases?

– How was the mobilization organized and what are the processes implemented to this end?

– What is the real impact of these forms of mobilization on political decision-making?

From a methodological point of view, this study uses two approaches (qualitative and quantitative).

## 2.2. Social networks and the question of their importance in the world, in Africa and in Côte d'Ivoire

In recent years, social media have been a real phenomenon on the Web. Millions of people around the world are using them with increasing frequency to the point of becoming "cybercitizens[5]". This has transformed society as a whole to the point of rethinking political participation. Political actors make it a privileged tool in the daily and global management of their political actions. To this end, the use of digital social networks in the 2008 US elections has convinced the most skeptical of its new status as a political communication tool. On this subject, Maria Mercanti-Guerin states:

> "The last US presidential election recently highlighted the importance of the Internet and its social networks in the success of a campaign. The links between the Internet and candidates' strategies, the challenges of the emergence of this

5 Citizens exercising their citizenship through digital tools.

new medium in political discourse have given rise to a new interest in political marketing."[6]

The development of new information technologies in Côte d'Ivoire started in 1995, with the advent of mobile telephony followed by the Internet. Côte d'Ivoire established its first Internet connection in 1996 with the Internet service provider (ISP) Africaonline.

Côte d'Ivoire is connected to the Internet on two international networks: the Leland node and SAT3. The entire Internet ecosystem in Côte d'Ivoire revolves around Côte d'Ivoire Télécom (CIT), the incumbent operator, present at all levels of the distribution chain. Arobase Telecom, acquired by the mobile operator MTN, has a fiber optic network throughout the city of Abidjan and some major cities. Both companies operate directly on the American submarine cable system SAT3[7].

In Côte d'Ivoire, there are two main ways to connect to the Internet: via the ADSL line, i.e. fixed or wireline telephone lines, or via mobile Internet provided by mobile operators who essentially use two technologies, namely 3G, soon to be 4G and Wi-Max[8] technology. Internet distribution to households and businesses is provided by five ISPs: YouMee, Afnet de MTN, Vipnet, Africaonline from Afrique Technologies, Alink from Atlantique Telecom and Aviso de Côte from Ivoire Télécom.

The advent of mobile Internet is a real opportunity for everyone to connect. Internet connectivity in Côte d'Ivoire has evolved, and now mobile operators (MTN-CI, Moov-CI and Orange-CI) offer Internet solutions in mobile situations. These operators use GPRS/EDGE, CDMA, 3G, 4G and Wi-Max technologies. These means of connection to the Internet can be categorized into three groups: USB Data Internet sticks[9], hubs[10] and Wi-Fi hotspots[11].

---

6 Analyse des réseaux sociaux et communautés en ligne: quelles applications en marketing ? [Online]. Available at: https://www.cairn.info/revue-management-et-avenir-2010-2-page-132.htm [Accessed July 24, 2014].

7 See on this subject *Côte d'Ivoire Économie*, June 2010.

8 The term "Wi-Max" refers to a wireless network standard, which is the abbreviation for Worldwide Interoperability for Microwave Access. Wi-Max allows data transmission with a theoretical throughput of up to 75 Mb/s over several tens of kilometers.

9 The USB key allows us to connect to the Internet using the network of telephone operators.

In addition, the Ivorian government is committed to democratizing Internet access through the "one citizen, one computer + one Internet connection" project implemented by the *Agence nationale du service universel des télécommunications-TIC* (ANSUT)[12].

Since the introduction of Internet technology, it has rapidly expanded in terms of habits and households. In 2000, the number of users was estimated at 20,000. This number increased from 119,526 in 2012 to 133,982 in 2013 according to the Observatory of the Telecommunications Regulatory Agency of Côte d'Ivoire (ARTCI). It should be noted that, until 2013, statistics did not take into account mobile network users.

Today, in light of the rapid evolution of mobile networks, particularly 3G and 4G, this number has changed considerably. The World Stats[13] website published for the year 2015 the figure of 5,230,000 Internet users in Côte d'Ivoire, for a penetration rate of 22.5%.

ARTCI, for its part, estimates that the total number of Internet subscribers in Côte d'Ivoire in the first quarter of 2015 has increased to 7,222,261, of which 7,155,583 are 3G subscribers. According to data from the Regulatory Authority, the overall Internet penetration rate is 31.65%.

At the time of writing this chapter, it is impossible for us to provide the number of mobile Internet users by operator.

Table 2.1 provides an overview of the number of mobile network subscribers by operator as of March 31, 2016.

Of the three mobile telephone operators in Côte d'Ivoire in 2016, according to the Côte d'Ivoire Telecommunications Agency, there were 11,253,361

10 A hub is a type of box used as network termination equipment, which an Internet Service Provider (ISP) provides to its subscribers for Internet access.

11 A Wi-Fi hotspot, or a Wi-Fi point, or a terminal, is a device that provides access to a Wi-Fi wireless network that allows mobile device users to connect to the Internet. The access thus provided may be free or for a fee for the user.

12 Available at: https://www.ansut.ci/web/projets/ [Accessed November 30, 2015].

13 Available at: www.internetworldstats.com, statistics based on data from the International Telecommunication Union (ITU) and WWW [Accessed November 30, 2015].

with Orange, 7,904,339 with MTN, 5,396,791 with MOOV or 2,4554,491 subscribers[14].

It should be added that the number of Internet subscribers in Côte d'Ivoire has increased since 2012, from 119,526 to 8,712,626 subscribers, an increase of 34.45%[15] according to ATCI.

| | Operators | | | | | |
|---|---|---|---|---|---|---|
| | **MTN Internet** | **Aviso** | **Vipnet** | **AlinkTelecom** | **Yoomee** | **Total** |
| **Access via wireline technology** | 558 | 56,645 | 2 020 | 41 | – | 59,264 |
| **Wireless via technology access** | 39,265 | 4,661 | 199 | 19 | 6,084 | 50,228 |
| **Total** | 39,823 | 61,306 | 2 219 | 60 | 6,084 | 109,492 |

**Table 2.1.** *Breakdown of the fixed Internet subscriber base by technology as of March 31, 2016 (source: ARTCI, March 31, 2016)*

## 2.3. Some message content broadcast on digital social networks

An interview with the publication director of the general information daily newspaper *L'Éléphant déchaîné*[16] allows us to establish the following facts:

> "On Sunday, March 23, 2014 at around 10pm, young Awa Fadiga was found unconscious, following an attack, by agents of the Adjamé-Mirador gendarmerie. She was then transported to the emergency department of the Cocody University Hospital by the Indenié fire brigade. On Tuesday, March 25, her family was informed of her death. The causes of this tragedy differ according to whether one is on the family's side or on the side of the authorities. The death occurred 'as a result of her very

14 ATCI sources (first quarter of 2016).

15 ATCI sources (March 31, 2016).

16 Biweekly magazine of the Ivorian written press.

> serious condition' according to the authorities. The family claims that the negligence of the health personnel towards their daughter is at the root of the tragedy because the victim was only treated when they arrived, i.e. on Monday 24 March at 13:00."

The death was difficult for Awa Fadiga's relatives to accept. It led to a mobilization to prevent this type of "negligence" from going unpunished and to highlight the failures of the health system in general, and that of Cocody University Hospital in particular so that there could be no "other Awas in hospitals". Very quickly, an Awa Fadiga group was created to organize the "revolt". Digital social networks, led by Facebook, were used to mobilize public opinion. Pages and communities were created for this purpose:

– the Facebook page created on March 26, 2014: "*Awa Fadiga, stop plus jamais ça*" (in Awa Fadiga's name, Stop, never again);

– the public group created on March 26, 2014: "*Pour Awa Fadiga, levons-nous pour que ça cesse* !" (For Awa Fadiga, let's rise so it never occurs again);

– the public group created on March 27, 2014: "*Pour Fadiga Awa plus jamais ça*" (in Awa Fadiga's name, never again);

– a Facebook profile created on March 27, 2014: "*Pour Fadiga Awa*" (for Awa Fadiga).

Very quickly, the tragic loss of the young girl created a buzz. Mobilization on the public space of free online expression around this social subject motivated decision-making at the political level. It forced the Ministry of Health and AIDS Control to issue a statement insisting that the victim received "the primary medical care that her condition required" and that the scanner found that she was suffering from "head trauma and acute respiratory distress". Subsequently, three managers of the Cocody University Hospital were disciplined. An investigation into the circumstances of the death of young Awa was opened by the public prosecutor. The "Awa Fadiga case" was also a subject for the Council of Ministers regarding the adoption of "urgent measures to restore emergency services in university hospitals and peripheral health facilities in the city of Abidjan set up as general hospitals (GH)"[17].

17 Publication of April 9, 2014, Facebook page "*Awa Fadiga, stop plus jamais ça*" [Accessed August 4, 2015].

## 2.4. Mobilization and process implemented: a need for freedom met by the Internet

The Internet is increasingly becoming an essential tool both for citizens in their daily lives and for public authorities in the management of political affairs and the relationship with its citizens. How did this tool become a symbol of freedom to the point of giving rise to words or concepts such as cyber-democracy, digital democracy and digital society?

Changes due to the Internet and the Web give rise to new words or concepts. Today, the virtual, the digital and the electronic reign. The digital space has developed very rapidly in a few years. The Web has become the essential key to this cyber society, just as the telephone and television have also profoundly transformed our societies.

The Internet is constantly and significantly changing relationships between individuals by opening up a wide range of possibilities in terms of communication, as well as more generally in terms of individual freedoms, private life and social relationships. The observation of freedom satisfied by the Internet is done at several levels:

– Creation of new actors: statistics show that citizens are increasingly connecting to the Internet to access services or to communicate. Citizens, by heavily using websites, have become cyber consumers, without really realizing it. According to Henri Oberdoff: "Internet users have gradually become actors of the Web. They provide their contributions to an alchemy of multitudes or a new collective intelligence. Democracy can only be nourished by it".

– Breaking down the boundaries between different technologies: there is no doubt that democracy is in the age of the Internet. A "great digital conversion" is at work. The networking of computers around the world, the acceleration of ADSL, wireless and mobile connections, the creation of the Web and the development of powerful search engines are examples of fundamental innovations that are being adopted very quickly by many users around the world. The main consequence of these various technical changes is the breaking down of borders between the various information, communication and telecommunication technologies.

– A virtual space of great freedom: it is easy to measure how the Internet and the Web can foster new spaces for citizens to exchange ideas so that they can contribute to the vitality of democratic expression and even democratic hope. The notion of space is adapted to this virtual approach to democracy in which citizens make fuller use of their freedoms of information, communication and expression. It is quite symptomatic to see the reappearance via the Web of the terms "forum" or "agora" to describe this collective takeover of modern communication tools, as if the extent of the territories no longer prevented the forms of direct democracy borrowed from the ancient cities.

– The emergence of new social practices: the development of websites has fostered social appropriation of the Internet. Sites have evolved within lockstep with the Web. If the first generation of the Web gave a relatively classic one-way communication, Web 2.0 allowed the development of interactivity. In this case, electronic agora and e-government become new possibilities for the functioning of democracy. This interactivity has also given rise to new social practices, for example, with digital social networks that foster the emergence of new relationships between individuals.

– The possibility of circumventing State controls: Internet use also pushes for democratic developments in countries that do not yet really practice it, taking into account the freedom of communication they induce. Even when governments try to control Internet practices, Internet users devise effective ways to circumvent these controls[18].

## 2.5. The real impact of mobilization on political action

The use of digital social networks to inform and mobilize one another, to organize demonstrations, to denounce injustices and to raise awareness among the Ivorian population is not new. Some events, which seem unimportant at first glance, have brought the social web to the point of making it a major topic for public authorities. Let us take the following examples:

– in March 2014, digital social networks made it possible to stay ahead of traditional media in dealing with the tragic disappearance of the model Awa Fadiga;

– digital social networks were being used as a framework for denouncing the phenomenon of child abduction and murder during the months of January and February 2015;

18 All these reflections are taken from the April 9, 2014 publication of the Facebook page "*Awa Fadiga, stop plus jamais ça*" [Accessed August 4, 2015].

– digital social networks allow the population to denounce violence, murder and theft attributed to “*les microbes*”[19];

– digital social networks have also been the framework for contesting the increase in electricity prices from March 2016. Without waiting for consumer associations, citizens have undertaken virtual protest actions through the social web.

Finally, digital social networks are also a forum for debate on political issues. By way of example, we can recall the question of the eligibility of candidates for the October 2015 presidential election: for days, it was at the heart of debates on social networks during the period preceding the validation of candidates by the Constitutional Council. The project to draft the new constitution is therefore increasingly becoming one of the topics of debate on the social web.

Therefore, it appears from events that in Côte d’Ivoire digital social networks make it possible to discuss important issues, generate public opinion, claim rights, organize collectively and defend the fundamental freedoms that every individual in the world is entitled to demand and be granted today. It seems appropriate, at this stage, to consider the opportunities that digital social networks offer for political participation in Côte d’Ivoire.

The socio-political situation prevailing in the country seems to encourage the initiative of such scientific reflection. In fact, Côte d’Ivoire has been experiencing a series of events for more than a decade that have seriously affected political life and social cohesion. In this context of negative peace, spaces for free expression are shrinking and some foundations of democracy are threatened. It is in such an atmosphere that the increasing use of digital social networks is observed.

The increasingly regular use of these online communication tools will gradually transform users into web actors, as well as into actors in political life. Increasingly, digital social networks are used to alert public opinion to certain social injustices experienced by some citizens.

---

19 “*Les Microbes*” is the nickname given to the gangs that have been spreading hysteria among the Ivorian population since the end of the post-election crisis. They are characterized by their age (usually minors), their weapons (knives) and their readiness to use violence against others.

We take as examples the phenomenon of *les microbes*, the phenomenon of child abduction and the Awa Fadiga case. In many cases, the enthusiasm generated around a particular topic on online social networks leads political authorities to take an interest in it. It seems that change is on the horizon. This is what Milad Doueihi (Oberdoff 2010, p. 8) calls the "great digital conversion"[20]. Like Tunisia and Egypt, democracy in Côte d'Ivoire is being combined with the digital age. Today, it is difficult if not illusory to build democracy without taking into account the growing importance of virtual public spaces in the daily lives of citizens.

## 2.6. Mobilization and messages

Mobilization and messages are carried via the media (social networks and traditional media) and by means outside the media field.

It is important to note the existence of the mobilization of traditional and non-media media, although both have experienced discontent manifested on social networks.

Mobilization via social networks was effective because of the signing of petitions and the creation of two dedicated platforms, notably on Facebook and Twitter, "*Les 200*" respectively created following the speech by the government's spokesperson, Bruno Nabagné Koné, Minister of Communications and the Digital Economy, on April 21, 2016:

> "Verbal scandal! I am Minister of Telecommunications and I know that there is only unnecessary noise on social networks. Only 200 Ivorians are pretending to be suffering, otherwise Côte d'Ivoire is doing well" (Ozhylly 2016), held on Wednesday, April 20, 2016 after the Council of Ministers.

These platforms have mainly conveyed messages of increasing citizens' awareness of the problems of the high cost of living, of calling for the massive diffusion of ideology, of gathering, of adopting dress codes as a sign of disapproval of government actions.

---

20 This expression refers to the way in which information and communication technologies, or electronics in particular, play a decisive role in the development of democracies and democracy.

This type of mobilization has also been reported in the traditional media, as well as by most members of civil society, including consumer associations, opinion leaders and the opposition (party leaders). In this case, Pascal Affi N'Guessan, leader of the main opposition party, is one of them. The President of the Ivorian Popular Front (FPI):

> "called on his supporters and Ivorians to 'mobilize' to put an end to the 'tragedy' of the power of Head of State Alassane Ouattara, faced with violent social demonstrations linked to the increase in electricity prices [...] Mr. Affi N'Guessan drew an uncompromising assessment of Mr. Ouattara's power, judging him 'politically and economically incompetent'. [...] Regarding the increase in electricity tariffs, the president of Mr. Gbagbo's party called on the government to 'listen to the voice of the people' by simply cancelling this increase to lower discontent."[21]

With regard to the mobilization on the ground, discontent was initially the result of the gathering of 200 people in front of the premises of the regional directorate of the CIE of Yamoussoukro, to demand the cancellation of the deadlines for payment of invoices issued in July[22]. Because of the speech by the government's spokesperson Bruno Koné, mobilization on the ground has been transposed from the digital space into demonstrations that have had an impact on society in general.

Thus, "strikes have agitated the public administration, particularly within the Treasury, in the customs services or in universities. The strikers demand that the State respect its commitments: the high cost of living, particularly with the rise in the cost of electricity and the reform of driving licences, are at the heart of Ivorians' discontent" (Mieu 2016).

In this public space, apparent anonymity facilitates speaking out and inactivity allows for "formal equality of participants in the political debate" (Mercklé 2004). Behind the avatars, pseudonyms can be anyone: a normal

---

21 SKO, Grogne sociale en Côte d'Ivoire: le président du parti de Gbagbo appelle à la mobilization des Ivoiriens [Online]. Available at: http://news.abidjan.net/h/594953.html.

22 À Yamoussoukro près de 200 personnes manifestent contre les "dates limites de paiement" [Online]. 2016. Available at: http://www.connectionivoirienne.net/119424/cote-divoire-electricite-a-yamoussoukro-pres-de-200-personnes-manifestent-contre-dates-limites-de-paiement [Accessed December 5, 2016].

citizen, a politician, an authority, etc. The important thing is that each member of the community can express his or her point of view, similarly to the "Awa Fadiga case".

## 2.7. Social networks, political actors and the exercise of power in Côte d'Ivoire

Today, for reasons of image and credibility, political actors are required to be present on social networks. They must be in the know and show that they can master technological tools. With the development of digital social networks, traditional media are losing their effectiveness; it is inconceivable to conduct a political campaign without using social media. Whether or not politicians like this communication tool and regardless of their technological culture, it is important to be present on social networks, whether on Facebook, Twitter, LinkedIn, etc. Realizing the importance of this new communication tool, the President of the National Assembly makes extensive[23] use of social networks. Aware of Facebook's growing influence, five days after the 2011 crisis, the President of the Assemblée nationale created his official Facebook page to communicate with Ivorians. He published constitutional declarations almost every day. As a result, most government members created a Facebook page.

Mobilization through social networks triggered the reaction and action of local authorities both with regard to the cost of electricity, the issue of permit renewal and the Awa Fadiga case. It is important to recall that, contrary to the problem of renewing the permit, the issue of increasing the price of electricity has led to discrimination and minimization in its propensity to mobilize on the part of the government's spokesman minister, Bruno Koné, Minister of Communications and the Digital Economy.

Apart from this ministerial effort, the social situations described in this chapter have all given rise to a strong reaction of opposition from local authorities. By way of illustration, we present the intervention of the two highest local authorities who are the President of the National Assembly and the President of the Country.

23 Classement des pages Facebook les plus populaires en Côte d'Ivoire , available at: https://lactuwebdedith.wordpress.com/2013/06/25/classement-des-pages-facebook-les-plus-populaires-en-cote-divoire/.

Mr. Soro Guillaume, President of the National Assembly of Côte d'Ivoire, first referred to the social discontent that is undermining the country. He also reiterated the main reason for this: the high cost of living observed by all socio-professional strata. In particular with the increase in electricity bills in households following the government's decision to increase and the renewal of driving licenses, which are required of users.

At the opening of the first ordinary session for the year 2016, he reassured citizens about the mobilization of leaders and particularly the imminent intervention of the President of the country on the issue:

> "We cannot pretend to ignore these voices, which are becoming more and more heard and question us about the cost of living and the opportunity of certain measures that our brave citizens do not always understand: rising electricity bills, recurrent disturbances in our universities, changes to the current version of the driving licence, failure to apply the texts imposing a ceiling on the amount of real estate securities to be paid to landlords, etc. The discontent observed here and there may indeed surprise us when we consider the progress made in five years by Côte d'Ivoire on the road to its economic, physical and moral reconstruction. Unfortunately, this is the downside of success. But in saying this, I also acknowledge that expectations are still high... Our citizens are demanding more inclusiveness, more solidarity. It is under this prism that we must understand and listen to the questions that are expressed in various forms and through various channels: press articles, SMS, emails, social networks, debates in markets and social clubs." (Kautcha 2016)

Alassane Ouattara, President of the country, made two major interventions. In the first, without taking any decision, he expressed his disapproval of the situation and mainly of the ministers in charge of the covering portfolios:

> "On Wednesday, at the Council of Ministers, the Ivorian President reportedly asked the government for an explanation of the social discontent in the country. The Council of Ministers on April 27 at the Abidjan's presidential palace was not easy for the members of the government. According to our information, Alassane Dramane Ouattara has asked for explanations about

> the social discontent that is currently stirring the country. He therefore questioned the ministers in charge of the dossiers, such as the Minister of Trade, the Minister of Transport and the Minister of Mines and Energy. The president was not happy and he let it be known." (Mieu 2016)

In his second intervention, during his speech to the nation on Labor Day on May 1, 2016, President Alassane Ouattara not only condemned, but also initiated measures to put an end to this social discontent.

As for the Ivorian Electricity Company (CIE), it has opened up the sector to competition, thus ending Bouygues' monopoly. He stated that:

> "With regard to electricity, it is true that in June 2015, the Government decided to set the electricity tariff at just 16% over three years. This decision was aimed at increasing the production, transmission and distribution of electricity and ensuring better access to electricity for citizens. The first phase of this 10% adjustment, effective July 1 2015, did not affect social rate subscribers, who represent 40% of all subscribers. The second 5% phase, applicable as of January 1, 2016, does not apply to social rate subscribers or small consumers, who represent 70% of subscribers for these two categories. So I didn't understand why our bills had gone up. After investigation, we found that the government's decision was not properly implemented and that some subscribers experienced a higher increase than initially expected, ranging up to 30 to 40% for some subscribers. Yes, there was an injustice in the increase that was made and that is unacceptable. I have therefore decided to cancel the January 2016 increase altogether. Invoices will be corrected and the overpayment will be returned to all subscribers. I therefore invite the CIE to redress this injustice… This situation calls us to consider the need to open up the water and electricity sectors to competition, as provided for in the electricity code. Yes, it is competition that will lower the price of electricity. I therefore call to all those who wish to invest in this sector." (La Rédaction 2016c)

Regarding the reform of driving licenses, President Alassane Ouattara said, still on the same date:

> "I also took a closer look at the operation to renew driving licences initiated as part of the reform of the transport sector. It is my belief that driver's licence renewal should not apply to all drivers. I have therefore decided to suspend the operation until the government makes proposals to me in the coming weeks on the conditions for continuing this operation [...] why renew a licence that is not yet ten years old?"

## 2.8. Conclusion

The impact of mobilization via social networks on the way social problems are solved, as noted in the previous section, does not involve any resolution through discussion.

Indeed, the resolution of the social problems relating to the increase in electricity prices and the question of driving licenses were effective by the unilateral decision of the President of the Republic.

When we consider the reasons that motivated the prompt reaction of the public authorities to the issues of rising electricity prices, the Fadiga case and the renewal of driving licenses, it must be noted that social networks can be considered an instrument for measuring the exercise of power in Côte d'Ivoire.

## 2.9. References

Bujold, C. and Gingras, M. (2000). *Choix professionnel et développement de carrière. Théories et recherches*. Gaëtan Morin, Montreal.

Canivez, P. (2013). *Qu'est-ce que l'action politique?* [Online]. Available at: http://iphilo.fr/2014/03/01/quest-ce-que-laction-politique [Accessed 5 December 2016].

Charbonneau, J. (2004a). Les réseaux sociaux: l'approche relationnelle. Atelier d'experts sur la mesure du capital social. Projet de recherche sur les politiques. Ottawa, June, 1–7.

Charbonneau, J. (2004b). Sociabilité et réseaux sociaux. Notes de cours. Institut national de recherche scientifique, Montreal.

Charbonneau, J. (2005). Associations en réseaux et capital social collectif: Bilan d'expériences et d'analyses. Report, Institut national de recherche scientifique, Montreal.

Dieth, A. (2016). Les réseaux sociaux, médium de la réconciliation en Côte d'Ivoire [Online]. Available at: https://blogs.mediapart.fr/alexis-dieth/blog/220516/les-reseaux-sociaux-medium-de-la-reconciliation-en-cote-d-ivoire.

Fogel, J.-F. and Patino, B. (2014). La condition numérique. Questions de communication [Online]. Available at: http://questionsdecommunication.revues.org/8841 [Accessed 23 December 2016].

Ghetty, C. and Chauvet, V. (2009). Le rôle informationnel joué par les communautés virtuelles: l'influence de la méta-information [Online]. *Revue internationale d'intelligence économique*. Available at: http://www.cairn.info/revue-internationale-d-intelligence-economique-2009-1-page-93.htm [Accessed 15 December 2016], 1, 93–105.

Kabla, H. (2014). Voici trois moyens pour construire une France numérique [Online]. Available at: https://www.huffingtonpost.fr/herve-kabla/cher-emmanuel-macron-voici-trois-manieres-dapprehender-le-numerique_b_5803980.html [Accessed 18 July 2016].

Kahofi, S. (2011). *Impact des réseaux sociaux en Côte d'Ivoire* [Online]. Available at: http://eburnietoday.mondoblog.org/2011/01/28/impact-des-reseaux-sociaux-en-cote-d%E2%80%99ivoire [Accessed 5 December 2016].

Kautcha, D. (2016). Côte d'Ivoire: Grogne sociale, Soro surfe sur la vague de la contestation. [Online]. Available at: http://koaci.com/cote-divoire-grogne-sociale-soro-surfe-vague-contestation-98071.ht [Accessed 5 December 2016].

Klen, M. (2010). *Les ravages de la désinformation*. Favre, Paris.

Kra, R. (2013). Communication et développement local: la contribution des technologies de l'information et de la communication (TIC) à la gouvernance locale [Online]. *La communication en question*. Available at: www.comenquestion.com [Accessed 25 October 2016], (1), 81–99.

La Rédaction (2016a). Côte d'Ivoire/Bruno Koné dresse "un bilan positif" de l'action politique du président Ouattara [Online]. Available at: www.News.abidjan.net/h/5999048.html [Accessed 5 December 2016].

La Rédaction (2016b). Côte-d'Ivoire électricité: À Yamoussoukro près de 200 personnes manifestent contre les "dates limites de paiement" [Online]. Available at: http://www.connectionivoirienne.net/119424/cote-divoire-electricite-a-yamoussoukro-pres-de-200-personnes-manifestent-contre-dates-limites-de-paiement [Accessed 5 December 2016].

La Rédaction (2016c). Grogne sociale en Côte d'Ivoire: La reculade de Ouattara après des échanges houleux avec les syndicats (1er mai) [Online]. Available at: http://www.connectionivoirienne.net/117921/grogne-sociale-cote-divoire-ouattara-prend-mesures-fortes-apres-echange-houleux-syndicats [Accessed 5 December 2016].

La Rédaction (2016d). Médias sociaux et contestation politique [Online]. Available at: http://www.zones-subversives.com/2015/03/medias-sociaux-et-contestation-politique.html [Accessed 5 December 2016].

Laurie, F. (2014). La liberté d'expression et les réseaux sociaux [Online]. Research report, Université d'Aix-Marseille. Available at: http://junon.univ-cezanne.fr/u3iredic/wp-content/uploads/2014/03/Les-r%C3%A9seaux-sociaux-et-la-libert%C3%A9-dexpression.pdf [Accessed 19 December 2016].

Mandour, D. (2013). Internet: nouvel outil de démocratie? Étude de cas de l'Égypte. Master's thesis, Université de Paris 1, Paris.

Mieu, B. (2016). Côte d'Ivoire: face à la grogne sociale, Ouattara en rogne contre ses ministres [Online]. Available at: http://www.jeuneafrique.com/321441/politique/cote-divoire-face-a-grogne-sociale-ouattara-rogne-contre-ministres [Accessed 5 December 2016].

Oberdorff, H. (2010). *La démocratie à l'ère numérique* [Online]. Presses universitaires de Grenoble, Grenoble. Available at: https://www.lgdj.fr/auteurs/henri-oberdorff.

Ozhylly, G. (2016). Scandale! Grogne sociale: quand Koné Bruno pense plaire à Ouattara face aux souffrances des Ivoiriens [Online]. Available at: http://www.ledebativoirien.net/2016/04/22/scandale-grogne-sociale-quand-kone-bruno-pense-plaire-a-ouattara-face-aux-souffrances-des-ivoiriens [Accessed 5 December 2016].

Séry, B. (2009). *Écrits pour la démocratie*. CERAP, Abidjan.

SKO. (2016). Grogne sociale en Côte d'Ivoire: le président du parti de Gbagbo appelle à la mobilisation des Ivoiriens [Online]. Available at: http://news.abidjan.net/h/594953.html.

Tété, E. (2011). Côte d'Ivoire: usage innovant des réseaux sociaux [Online]. Available at: http://cursus.edu/article/10071/cote-ivoire-usage-innovant-des-reseaux/#.WErWQPnhDIU [Accessed 5 December 2016].

Zamma, N. (2012). Réseaux sociaux numériques: Essai de catégorisation et cartographie des controverses [Online]. PhD thesis, Rennes 2, Rennes. Available at: www.halshs.archives-ouvertes.fr/tel-00687906/document [Accessed 16 November 2016].

# 3

# Mobilization for Collective Action: The Contribution of Digital Social Networks to Political Engagement

For a long time, people were served by traditional media (print, radio, television), which made it possible to disseminate and propagate information all over the world. Over time, humanity has generated new information devices that are translated into digital social networks (DSNs) with the arrival of computers and the Internet. Due to these networks, the world is free from geographical and temporal constraints; bringing Man into the world of modern science, "where time, freed from time, is converted into a space of pure information no longer requiring human corporeality", to use David Le Breton's (1999) words. We are in the era of this period of change and acceleration in all directions, of sharing and collaboration, of discourse, of activism and revolution, of emergence and disappearance.

Today, social networks have multiplied all over the world and are positioning themselves as niches for collective mobilization. After email, online libraries, music or film downloads, forums and blogs, we now have DSNs and with them, the changes they have brought about and the new communication possibilities they offer. We are living in an era of change with communication and access to information that previously depended on the place and time and are now "deterritorialized" and "detemporalized" (Zammar 2012, p. 18). This success is linked to the possibilities offered by the Internet.

---

Chapter written by Gbati NAPO.

Worldwide, there were 3.773 billion Internet users in 2017 with a 50% penetration rate for a world population of 7.476 billion people[1]. In Africa, the number was 362 million Internet users with a 29% penetration rate for a population of 1.231 billion. In 2010, Africa had nearly 6,000 of the most connected Arabic-speaking bloggers in the Middle East. One of the We are Social agency's reports identified 170 million social network users in Africa with a 14% penetration rate. The Facebook platform, created in 2004, is one of the most widely used networks and now has more than a billion active visitors on a daily basis. The number of users surged in 2016: 146.6 million African Internet users have a Facebook account according to a study conducted by the Tunisian company Medianet[2]. The main figures updated by *Sen Marketing* (February 2017) are as follows: Facebook: 168 million registered users; Instagram: 31 million registered users; LinkedIn: 24 million registered users. There are more and more Internet users due to the drop in cell phone prices (Android smartphones), as well as thanks to the relative improvement of telecom networks on the African continent. Thus, in 2006, the World Bank declared that "the information revolution […] offers Africa a dramatic opportunity to leapfrog into the future, to break decades of stagnation and decline". And Marie-Soleil Frère and Fanny Heuzé (2007, p. 88) rightly speak of "a new key for development".

According to a report by the digital agencies We Are Social and Hootsuite, Africa recorded a 12% growth in active social network users in 2017, bringing the number of users to 191 million across the continent. Of these, about 172 million use only two Facebook-owned platforms: WhatsApp and Messenger. Since January 2017, Internet penetration in Africa has increased by more than 20%, reaching more than 73 million additional people[3].

In just a few years, digital social networks have become one of the main uses of the Web on the African continent. "In Côte d'Ivoire in particular, several traditional associations operating in networks have transposed their

1 Digital in 2017 global overview [Online]. Report, January 25. Available at: https://wearesocial.com/special-reports/digital-in-2017-global-overview [Accessed March 22, 2018].

2 Available at: http://www.rfi.fr/technologies/20161124-internetreseaux-sociaux-citoyennete-numerique-afrique [Accessed March 23, 2018].

3 Available at: https://afriquenligne.fr/2018/02/07/utilisation-dinternet-afrique [Accessed 15 March 2018].

activities into digital technologies" (Dagnogo 2018). Each individual seems to be a *homo communicans*. It is clear that these networks have become some of the most consumed products on the Internet. This observation is widespread not only in the West, but also in Africa. It is in this context of technological change that economic, political, social and social movements have become intertwined.

On the African continent, people tend to live in precarious situations which leads them to seek the improvement of their living conditions through collective action. They demand social change, constitutional and institutional reforms. In countries, these demands are expressed as requests for political alternation, restoration of rights, transparency and good governance or simply a better life. These collective movements in the name of freedom of expression are guided by an orientation towards a democratic ideal that revolves around the values of social justice, equal rights, dignity, freedom, transparency in the management of public affairs and the ability of individuals and groups to determine themselves. In other words, collective action aimed at transforming society. These movements have found a favorable response through digital social networks because they offer the possibility of a wide dissemination of information and therefore contribute to collective mobilization and adherence to the principles of advocacy. In this context, the problem that we will investigate is the following: how are collective mobilizations of citizens for socio-political demands made possible in the majority of African societies in the digital age through social networks?

This study is based on the following approach. The analysis is intended to be qualitative (Mucchielli 1994) because it involves clearly understanding a human fact whose social value generally escapes quantification. It favors the empirical research methods in communication proposed by Gérard Dérèze (2009), which is built on the observation and examination of the discourse and messages collected. Documentary research and consultation of some discussion platforms are of paramount importance. The choice of the information collection methods employed in this study is mainly justified by the fact that directories of mobilization actions always have their argumentative counterparts (Rennes 2011). By structuring the article in two parts, we intend to present the current situation of social networks and the theoretical positioning framework first, and the results of the research second.

## 3.1. State of play and theoretical framework

### 3.1.1. *Digital social networks (DSNs): current situation*

Since the concept of social networking was introduced by Barnes in 1954 and the emergence of Web 2.0 in 2004, it has continued to attract the interest of researchers. With the advent of the Internet starting in the 1990s, the generation of Web 2.0 users is referred to as Generation Y (Dudezert *et al.* 2008). Digital social networks (DSNs) allow the development of what is called collective intelligence (Malone *et al.* 2009). According to Rosanvallon (2006), there is a current trend of "counter-power" that originates in an ambitious attempt by the people to achieve social appropriation of power in a form of participation that the author describes as "unconventional". This is precisely why Reguer claims that DSNs are "a new form of political propaganda" (Reguer 2009, p. 10). For his part, Zammar (2012) points out in his analyses that DSNs insert reality back into the virtual world and the latter modifies reality in return. Michel Serres, quoted by J. Lespinas, states that "from a metric space with a measurable physical distance level, we have entered a new digital living space that involves the removal of spatio-temporal barriers, the mutation of individual identification and new interaction processes" (Lespinas *et al.* 2013).

Attalide announced that democracy will be electronic and "politics will disappear" because of networks, while Castells states that "networks destroy state control over society and the economy". In this perspective, the State does not escape the dissolving digitalization of the network. The Internet becomes the ideal model of the libertarian network, unlike the state network model intended for centralized control. In addition, the "anti-hierarchical" network becomes synonymous with self-organization and equality, bringing Internet users together in virtual communities. Michael J. Jensen, Eva Anduiza and Laia Jorba (2012, p. 4) expand on this:

> "Digital media has facilitated the creation of new modes of political participation that have not existed before: individuals can use the Web to post their views on blogs or forum comments, upload political videos to YouTube, or join political groups on social networks."

The notions of *e-democracy, cyber-democracy* and *hyper-democracy* have evolved to give rise to the possibility of creating the democracy-Internet couple.

The militant uses of digital media technologies will be felt more and more (Landry 2012). Indeed, the "crisis movements" or certain Take the Square-type actions, carried out during the Arab Springs (Tunisia, Egypt) or in Turkey (occupation of Taksim Square and Gezi Park in Istanbul), have enabled a strong collective mobilization through the use of the DSNs. These actions are described by Romain Badouard (2013, p. 93) as "keyboard mobilizations".

The report of the Centre for Interdisciplinary Research on Communication, Information and Society (CRICS 2014) highlights two practices: alternative media practices (call for mobilization, tactical communication, cyberterrorism, public denunciation, electronic lobbying, reverse surveillance) with three types of control in order to restrict them:

– technological control (created upstream by software manufacturers and developers);

– judicial review (through the development or implementation of a regulatory regime that criminalizes or penalizes what are considered illegitimate uses of digital media technologies);

– state control (through surveillance and monitoring, harassment and repression).

On the other hand, there are practices that resist the governance of digital media technologies. These practices include circumvention of censorship, calls for mobilization, boycotts, legal challenges, political lobbying, electronic civil disobedience, etc.

In short, the social struggle with digital media technologies and for better governance of democratic states is part of a broader perspective of the fight for the democratization of societies (Landry 2012).

### 3.1.2. *Theoretical positioning of research*

The theory of social networks developed by Mark Granovetter (1975) focuses on relational structures. The latter, in his foundational article "The Strength of Weak Ties", considers that individuals maintain a myriad of relationships which are of a different nature and distinct contributions in terms of benefits. Granovetter (1975) classifies social network links into two different types. "Weak ties" that are represented by acquaintances, occasional acquaintances or friends of friends. These links are a source of useful and non-redundant information because they are relationships established between actors in different contexts or environments. Putnam describes these advantages of weak ties as bridging social capital. The second type is that of "strong ties", which refer to relationships that naturally occur between friends or family members. These strong relationships, described by Putnam as bonding social capital, are essentially considered a source of emotional and social support and tangible benefits such as financial loans or assistance of any kind. For this work, the social network theory developed by Granovetter (1975) allows us to analyze how messages calling for participation in public demonstrations reach unknown people or weak links (Facebook pages or WhatsApp groups) on the physical level to rally them to the cause defended by the initiators of collective actions. The data collected and analyzed are presented in the second part of our study.

## 3.2. Research results

### 3.2.1. *Types and processes of collective action: the different types*

There are several types of collective actions. Depending on the pressure one wants to put on an opponent, or the intensity of the fight, the modes used are not the same. Table 3.1 shows some examples of group actions.

| The most well-known | The most "alternative" ones |
|---|---|
| Placard demonstrations<br>Fax campaigns, petitions<br>Letter of support campaigns<br>Bridge and road blockages<br>Strikes<br>Marches<br>Popular uprisings<br>Riots<br>Shoe demonstrations | Non-violent direct actions<br>Invisible theater<br>Boycotts<br>Passive resistance<br>Urban guerrillas<br>Civil wars<br>Bands on foreheads/arms<br>Secession |

**Table 3.1.** *Illustration of known group actions with the most alternative ones*[4]

4 Source: Adapted from the summary sheet Action Collective, Regroupement des auberges du cœur, p. 2 (2009). Available at: http://www.aubergesducoeur.org/files/Fiche%20synth%C3%A8se%20-%20Action%20collective%20VF.pdf.

Table 3.1 presents two types of collective actions. The first category represents those that are known and more practiced, i.e. demonstrations, strikes, marches and more recently the shoe demonstration that was observed at COP21 (where some people left their shoes on the road to go about their business), or the protests of the Yellow Vests in France. The second category which we are witnessing represents collective actions that are not necessarily known in some places, for example, bands on the forehead or arms to express dissatisfaction, invisible theater (e.g. shutting off the light in the factory and acting as if nothing had happened), or secession. The goal of these types of mobilization is to demand an improvement in social life, the well-being of people and respect for the values of democracy.

### 3.2.2. *Collective action process*

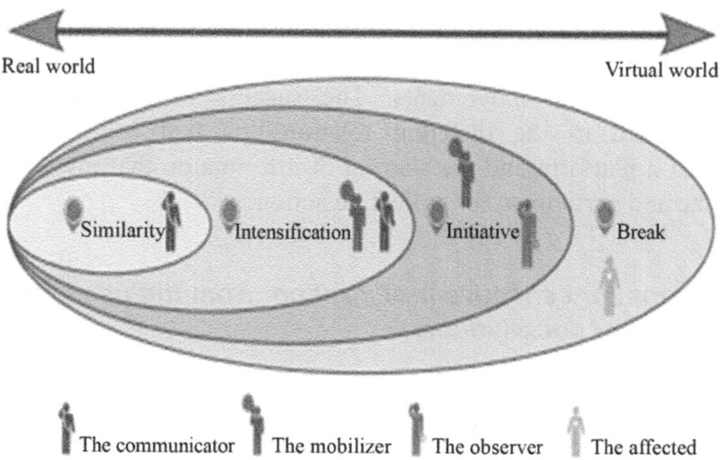


**Figure 3.1.** *Example of a collective mobilization process5. For a color version of the figures in this book, see www.iste.co.uk/atchoua/digital1.zip*

Many authors have described what the group mobilization process is to them. We focus on two processes: the first is illustrated in Figure 3.1 and the second is based on an analysis of field data. Starting from the diagram, we observe that a communicator alerts people to a problem situation, a frustration,

5 Source: Squalli, H. (2012). "Vie virtuelle, vie réelle: mêmes trajectoires?", *Usages et usagers des médias sociaux*. Economia, 14, 33. Available at: http://economia.ma/sites/default/files/ECONOMIA%20N%2014%20print%20web.pdf.

a sense of dissatisfaction. Using the power of social networks, they succeed in amplifying the results of their analysis of the situation and their frustration. They couple their message with a call to act. The call constitutes an initiative that will break with reality as shown in Figure 3.1. While the mobilization took place in the virtual world, the materialization of collective action will take place in real space.

Other analyses present collective action as follows. Attention makes it possible to realize that a situation is not right and that it is urgent to initiate micro-interaction with others by sharing information with them. Then, the mobilizer collects opinions regarding the initiative of call to demonstrate as well as the contribution of each other. Once this has been done, they can congregate around the demonstration idea and the mobilizer can thus create enthusiasm around the call to demonstrate.

These two processes make it possible to understand how social actors, virtually unknown, manage to find themselves in a real physical space to participate in protest movements. The success of this mobilization is therefore linked to the different relationships that bring together the members of a platform and the sharing of information that invites everyone to mobilize and participate for collective action.

### 3.2.3. *Actions for collective mobilization: from the use of traditional media to digital social media*

The introduction of mass media in postmodern societies has contributed to a profound change in the ways in which governments and the governed communicate. Before the advent of DSNs, collective mobilizations were made through leaflets, banners, griots, gongs, writing on buildings, graffiti, word of mouth, radio messages, print media, trailers, door-to-door, meetings, etc. With these means of communication, the calls to demonstrate take time, remain tedious and require considerable physical, human and financial resources.

With DSNs, collective mobilizations are done differently and more quickly, requiring less resources or expenditure. The emergence of new practices generated by DSNs is grafted onto the past, onto routines making them cultural survival that persist and continue to be transmitted well beyond their appearance (Mallein and Toussaint 1994, p. 317).

The public's excessive use of social media (blogs, platforms, social networks, networking, content sharing, collaboration, etc.) as a source of information suggests that they are a new generation of communication channels competing with traditional media. Social media have thus become new channels for information, political debate and mobilization, advocacy, denunciation or even manipulation. Thanks to ICTs, people are beginning to freely organize outside the old channels (radio, print media, television) controlled by politicians. The so-called citizen mobilizations, characteristic of sovereign people, are gaining momentum (Tonyémé 2013, p. 57). ICTs have ushered in a new era in communication and, by extension, in organization and mobilization. They have transformed the practices of citizen participation and social demands, offering more opportunities and discussion in a space free of constraints. Considering the penetration rates and the increasing number of Internet users recorded in recent years on the African continent, it is clear that DSNs have gained ground in Africa with public support for the practices related to these technologies.

### 3.2.4. *Broadening public debate on DSNs and mobilizing for socio-political demands*

Different forms of citizen mobilization have been observed during this decade to support or participate in collective action related to the attainment of freedom of expression in today's postmodern societies. The advent of the Internet and DSNs have provided opportunities for citizens to express themselves in a variety of ways through contradictory debates. Information inviting popular mobilization for socio-political demands is disseminated and relayed by social networks.

### 3.2.5. *DSNs and broadening public debate in a context of democratic renewal*

For a few decades, the Internet has been nurturing the hope of a democratic revitalization. The mass use of network sites has extended the public debate to all issues of socio-political, economic and cultural life. On the one hand, elected officials, trade unionists, civil servants and media professionals have a broad voice in public affairs. On the other hand, citizens discuss political life in confined spaces (home, café, associations, etc.) (Beuscart *et al.* 2016, p. 95).

To what extent has the Internet, via DSNs, allowed for a broader public debate? As long as everyone can visit sites or exchange ideas online with people under different conditions that share similar orientations and convictions on a given subject, the debate is broadened without barriers and brings people closer together in space. For DiMaggio *et al.* (2001), the Internet significantly reduces entry barriers and the costs of participation in the electronic public sphere, making it easier for people to speak out than in traditional media. Most online discussion spaces (discussion groups, forums, blogs, social networks) are based on a principle of free publication. As a result, opinions and ideas considered inappropriate by traditional media can easily be expressed on the Web. Citizens most often invest discussion forums anonymously. The Web plays the role of a veil which, by suspending the identities of each person, gives rise to alternative statements in a context of general inter-knowledge and pressure to reach consensus (Beuscart *et al.*, p. 105).

In a context of democratic renewal, the rise of the Internet therefore contributes to broadening the public debate, although the forums for such broadening vary from one context to another. But the Internet has not only been used to disseminate information and discuss city affairs. It has also been invested in by political and civil society actors to renew forms of civic, political and citizen engagement.

This engagement makes it possible to provide civic and citizen education, to create and consolidate citizenship in the individual, which requires a set of skills and a language tool to participate in political and social life. This allows members of society to be prepared for democratic culture, develop a sense of participation in debates and the formulation of free opinions. This participation in social life confers a certain legitimacy on members of the society that makes them citizens (Napo 2013, p. 8).

ICTs, with the use of DSNs, seem to give a new impetus to the participation in socio-political life in a democratic context where citizens more are free to express their thoughts and opinions.

### 3.2.6. *DSNs and mobilization for political participation: elections*

For Jan Koum[6], the founder of WhatsApp, "simply, two centuries ago, we fought with guns, before that, with bows and arrows and now we fight with new technologies". DSNs, via the Internet, have raised hopes for engagement that not only reaches a broader segment of the population, but also is a means of mobilizing and actively participating in advocacy.

In an electoral context over the past two decades, several major mobilizations have taken advantage of the means of action offered by DSNs via the Internet to mobilize the electorate. One example is France in 2005 during the French referendum for the European Constitutional Treaty, where many blogs appeared to participate in the public debate online. Political parties use the Internet to coordinate the actions of their supporters, as was the case with the MyBarackObama platform model launched in 2008, during the US presidential elections. Barack Obama's victory in this election is a perfect example of how the Internet is being used to support a project to streamline campaign efforts (Beuscart *et al.* 2016, pp. 93–110). For its part, D. Kreiss (2013), cited by Beuscart *et al.*, (2016) shows how crucial the IT infrastructure put in place by Obama's team was to victory.

From this example, given the decline in voter turnout and the political apathy that can be observed everywhere, we can state that the central virtue of online technologies is to align the efforts of activists to make them more effective in order to contribute to the victory of their candidate. The metamorphosis of the communication space has resulted in profound changes in political communication techniques and tools. It is no longer only poster campaigns, radio broadcasts and television debates that mobilize people, influence or alter their electoral decisions, but also new media with the advent of the Internet and mobile phones. Today, social networks and mobile phones are powerful tools for political communication and electoral mobilization. Jacques Gerstlé (2008, p. 208) is probably right when he states that "among the various technologies used in election campaigns, the use of Internet-related practices is typical of the post-modern era".

6 Interview given to *Slon* magazine on February 27, 2014, translated by Manon Masset.

### 3.2.7. *DSNs and collective actions: examples of political engagement in Africa*

With the many protest movements spreading across the continent, the WhatsApp platform has played a key role in the flow of information. It is notably on this platform that activists turned to sound the alarm and thus avoid police violence during the distribution of leaflets.

#### 3.2.7.1. *Events of the Arab Spring*

December 2010 after Mohamed Bouazizi's self-immolation by fire, citizens bypassed censorship and control of the media established by the Tunisian government by using social networks to relay information within and outside their country. Indeed, the Tunisian revolution surprised everyone with a new phenomenon (Brun 2011). President Ben Ali's regime fell under popular pressure on January 14, 2011. Social media was then celebrated for enabling this revolution.

This Arab Spring, or what some intellectuals have agreed to call the Revolution 2.0 or "the Facebook Revolution", has highlighted the role played by digital social networks both in terms of exchanges and in terms of coordinating joint actions. Facebook, YouTube and Twitter have played a decisive role in the Tunisian revolution (Lecomte 2011), as well as in the Egyptian and Libyan revolutions. If we just had to stop at knowledge, it is clear that the audience of people who have heard of the movements would be reduced. This is precisely what Granovetter (1975) sought to demonstrate by opposing strong and weak ties.

#### 3.2.7.2. *The power of social networks demonstrated during the crisis in Burkina Faso*

In Burkina Faso, the snowball effect of social network use has led to significant social change after contributing to the removal of the former Blaise Compaoré regime in 2011[7]. Indeed, Facebook had been stormed by young people, who used it for calls to demonstrate and take to the streets. These calls have facilitated citizen mobilization and contributed to the circulation of information.

---

7 Available at: http://www.africa24tv.com/fr/burkina-faso-lepouvoir-des-reseaux-sociaux [Accessed March 23, 2013].

### 3.2.7.3. *Citizen mobilization via DSNs in Côte d'Ivoire*

Gnéré Blama Dagnogo wrote that in Côte d'Ivoire, digital social networks have played a major interactive role in citizen mobilization, particularly during the presidential elections of 2010 and 2015 (Dagnogo 2018, p. 2). During the crisis, the dissemination of information took several paths, especially through blogs and social networks such as Facebook (with clear positions depending on the trend defended). Images of some cities (Yopougon, Abobo, Bouaké or Cocody) were instantly sent around the world via the DSN. In each camp, the official Facebook pages (of Ouattara and Gbagbo) were flooded with comments in favor of one or the other candidate[8].

### 3.2.7.4. *Kémi Séba and the anti-CFA front*

After burning a CFA franc note on August 19, 2017 in a public square in Dakar, in protest against the lack of monetary sovereignty affecting countries using this currency, Kémi Séba was expelled from Senegal in September 2017. In October 2017, he moved to Benin. Kémi Séba said: "France will not be quiet until it abandons Africa", on April 8 at 3 pm, when the anti-CFA front launched a new non-violent international mobilization in Paris (Place des Invalides), according to his official YouTube page which has 6,446 subscribers[9]. His NGO had called for a huge mobilization on September 16 in the capitals of French-speaking Africa and the diaspora against monetary servitude[10]. This action was amplified by some songwriters who have made numerous calls for mobilization to win this difficult battle, which is of less concern to the governments of the countries concerned.

### 3.2.7.5. *The* "Y'en a marre" *movement*

The actions of the "*Y'en a marre*" (enough is enough) movement have been remarkable and have contributed to the departure of former President Abdoulaye Wade from power in Senegal. The protests conducted over several months used channels such as Facebook and YouTube where many videos were posted.

---

8 Available at: www.rfi.fr/afrique/201110228-crise-ivoirienne-vit-facebook-twitter [Accessed April 7, 2019].

9 Available at: https://www.youtube.com/channel/UCZAtrvoju_bYM3fEOes7AEA/videos.

10 Available at: https://www.cotedivoire.news/benin/20160-lutte-anti-cfa-kemi-seba-annonce-manifestations-afrique-diaspora.html [Accessed April 4, 2018].

#### 3.2.7.6. *The case of Togo: an illustration from political and activist actors*

Like many countries, Togo is suffering the full impact of the power of social networks. Today, political actors, activists and professional organizations have a Twitter account and a Facebook page. No one wants to stay on the sidelines of the computerization of the society we have been witnessing since the Web 2.0 years. We now give some examples of the appropriation of social networks aimed at mobilizing for group movements.

#### 3.2.7.7. *Socio-political crisis: August 19, 2017, or the new momentum of the protests against the regime in Togo*

Activist Bill Emile Davolk, whose subscriber base is 72,404[11], posted a message on his YouTube channel calling for a demonstration that he entitled: "Togo: Saturday, August 19, 2017, a date, a unique opportunity to end the Gnassingbé dictatorship." This message received 6,190 views. This call for the mobilization of all social strata expressed frustration with the regime and the plundering of the country's resources by a minority, as can be read in the post. The movement, like the Burkinabe revolution, hoped for the removal of Faure Gnassingbé' from power. Another video from the same activist said: "Togo: Let's all go out massively in the street on August 19, 2017. Our fatherland is calling us!" with 9,133 views; the posting said: "Call from the committed Professor Naboudja Bouraima for a patriotic start for all Togolese citizens on August 19." Togosite, an online newspaper, writes that "thousands of demonstrators found themselves on the streets in Lomé, as well as in the four other indicated inland cities, namely: Kara, Sokodé, Bafilo and Anié, and in many other foreign countries, to the call of TIKPI and the PNP"[12].

If analysts of Togo's socio-political situation agree that August 19 was a success with the synchronized marches, it is also and above all due to the power of gathering and mobilization offered by DSNs. DSNs made it possible to rally people who were not in the same localities, nor the same families, let alone the same political parties, to the cause. This was demonstrated by Granovetter (1975) in his theory of social networks.

---

11 Channel consulted on April 4, 2018. On its YouTube channel, the number of subscribers is à 72 404.

12 Available at: www.togosite.com-index-php-togo-8262-togo-pari-reussi-pour-tikpi-atchadam-et-le-pnp.

### 3.2.7.8. *Action by a Togolese diaspora activist*

Farida Nabourema's official YouTube channel has 4,327 subscribers. She relays the dates and places of gatherings in Togo and in the diaspora to "set fire to the buttocks", as she says. The IciLome website posted an article on December 8, 2014 entitled "Nabourema Farida, that little girl who scares the powerful UNIR", which has been read 39,713 times.

On Farida's blog, we can read several critical articles about UNIR representatives and government members. Among other titles, it reads: "Dear Togolese leaders, welcome to modern times", or "Call Togolese parliamentarians to demand reforms". "It must be said, without risk of being mistaken, that Farida is not one of those who admire the RPT/UNIR regime. On her blog, she alone surpasses Arc-en-ciel and CAP 2015 combined," writes the online media organization *IciLome*[13]. She writes on her blog that "in the human species, oppression is the weapon of predators"[14]. YouTube statistics show that she registered on March 15, 2009 and received a total of 549,938 views for 20 of her videos[15].

All these actions observed throughout the African continent allow us to state that, recently, due to the development of ICTs and DSNs, citizens have returned to the center of political action. They participate in political debates and can intervene directly in the management of their city without requiring the permission of politicians (Tonyémé 2013, p. 50). Paraphrasing Michle de Certeau, we can say that the increasing use of ICTs and DSNs to participate in the political debate is a revealing sign of the art of making digital users. These various mobilizations show that individuals play an active role in the adoption of technology (Alzouma 2014). They adapt themselves and also technology to the constraints and needs of their environment, demonstrate ingenuity and creativity and are no longer passive consumers.

## 3.3. Discussion

Diderot, Voltaire, like the other philosophers of Les Lumières, dreamed of a universal system to spread knowledge in order to fight ignorance.

13 Available at: http://news.icilome.com/?idnews=797735 [Accessed April 6, 2018].

14 Available at: http://www.faridanabourema.org/ [Accessed April 7, 2018].

15 Available at: https://www.youtube.com/user/fanaba19/videos?view=0low=grid [Accessed April 4, 2018].

This desire is growing with the Internet and social networks. The latter have seen the emergence of a category of Internet users engaged in participation in global public life. Whether they are activists, bloggers, WhatsAppers or tweeters, they succeed in influencing public action through the use of DSNs where there is freedom of expression and everyone seems to have the opportunity to express themselves without being politically correct. According to Eric Barendt's theorization (1985), the main challenges of freedom of expression are the search for truth and the consolidation of democracy. In this logic, the achievement of freedom of expression would result in the best possible society. In the context of this analysis, the objective targeted by Internet users through mobilization consists of providing information on a problem situation, taking initiatives and mobilizing for participation aimed at improving living conditions and therefore respect for the principles of democracy.

The Internet and DSNs appear as "a channel of civic expression that would upset the principles and justifications of representative democracy" (Vedel 2003, p. 81). It is a modern agora that stimulates, democratizes and enriches public debate through discussion forums, messaging and other plural spaces of expression. Of course, DSNs are not at the root of mobilizations for collective action. It is necessary to have the support and active participation of citizens. They are drivers that help disseminate information and relay it on a large scale. It is therefore essential to recognize that "if ICTs do not create revolutions, they can amplify them by establishing communication between citizens, a communication that cannot be controlled by the state apparatus" (Tonyémé 2013, p. 58).

If the Internet has increasingly established itself as the "fifth power", to use Thierry Crouzet's expression in societies with advanced democracy, it is because it is seen as a real counter-power capable of monitoring and blocking the multifaceted abuse of macro-powers (central or state power) as well as micro-powers (sectoral powers) spread throughout society. This power of the Internet via social networks is tending to take hold on the African continent with the multiplication of calls for mobilization observed in recent years.

And like any medal has its flipside, the use of social networks also has negative consequences. Identity theft, misuse of photos or images of others, defamation and cybercrime are, among other things, daily realities. The Cambridge Analytica case, a British firm that allegedly exploited, through backdoors, the personal data of Facebook network users to support Donald Trump's campaign, is in the media spotlight. With DSNs, misinformation, which is no longer a new phenomenon, is spreading further into the social environment. Fake news is spreading rapidly through digital networks.

## 3.4. Conclusion

Clearly, in the past and well before the use of social networks such as Facebook, Twitter, WhatsApp and others, there have been revolts, revolutions, social movements and mobilizations. But the management of new forms of engagement and mobilization for collective action now highlights the important role of these DSNs in this process. McLuhan's global village is no longer a utopia or a prophecy. The appropriation of ICT by citizens puts them back at the center of social and political action. While it is widely recognized that DSNs have become a place of identity expression, maintenance of social ties and information, they are also recognized as spaces for public debate and political and citizen engagement. The digital revolution and the multiplication of social networks are emerging as a new mode of global information exchange based on unrestricted sharing. ICTs have fostered communication, information sharing and democratized speech. Consequently, the mobilization and participation of citizens in collective action for a social claim is facilitated by these digital technologies.

Everyone agrees that the socio-political changes that have occurred in some Maghreb countries are the result of a strong collective mobilization with the help of DSNs. However, experience shows that the use of social networks for engagement or mobilization cannot be effective without a tie to the field and society. While we often speak of Arab revolutions (Tunisia, January 14, 2011 or Egypt, January 15, 2011), in particular the Tunisian revolution, as e-revolutions and if social networks have played an undeniable (as well as ambivalent) role, these are above all social movements marked by demonstrations, blockages, physical and where civil society action has been decisive. The use of DSNs is expanding today in other African countries and they are part of the strategies of social movements (Burkina Faso, Côte d'Ivoire, Togo, etc.). It is a strategy that undoubtedly has a future, given the ever-increasing number of Internet users and mobile phone subscribers.

But, in any case, the virtual space in which DSNs operate cannot replace physical space, as collective actions always manifest themselves and always end in physical space.

## 3.5. References

Alzouma, G. (2014). Faire de nécessité vertu : inégalités sociales et modes d'appropriation innovants du téléphone mobile en Afrique [Online]. *Revue française des sciences de l'information et de la communication*. Available at: https://journals.openedition.org/rfsic/946.

Attalide, J. and Castells, M. (1998). Supplément "Multimédia" du quotidien. *Libération*, 12 June.

Badouard, R. (2013). Les mobilisations de clavier. Le lien hypertexte comme ressource des actions collectives en ligne. *Réseaux*, 5(18).

Banque mondiale (1995). Rapport sur le développement d'Internet, March.

Beuscart, J.-S. *et al.* (2016). *La sociologie d'Internet*. Armand Colin, Paris.

Brun, F. (2011). La révolution en Tunisie. *Multitudes*, 1(44), 22–25.

Dagnogo, G.B. (2018). Du réseau social traditionnel au réseau social numérique: pistes de réflexion pour une éducation aux médias sociaux numériques en Côte d'Ivoire [Online]. *Revue française des sciences de l'information et de la communication*. Available at: https://journals.openedition.org/rfsic/3495.

DiMaggio, P. *et al.* (2001). Social implications of the Internet. *Annual Review of Sociology*, 27(1), 307–336.

Dudezert, A., Boughzala, I. and Mounoud, E. (2008). Comment intégrer la génération Millénial à l'entreprise?. In *Management: Enjeux de demain*, Pras, B. (ed.). Vuibert, Paris, 323–334.

Gerstlé, J. (2008). *La communication politique*. Armand Colin, Paris.

Granovetter, M. (1975). The Strength of Weak Ties. *The American Journal of Sociology*, 78(6), 1360–1380.

Jensen, M.J., Jorba, L. and Anduiza, E. (2012). Introduction. In *Digital Media and Political Engagement Worldwide*, Jensen, M.J., Jorba, L. and Anduiza, E. (eds). Cambridge University Press, Cambridge, 1–15.

Landry, N. (2012). Mouvements sociaux, technologies médiatiques et pouvoir. In *Penser les médias sociaux. Enjeux pour la communication*, Proulx, S., Millette, M. and Heaton, L. (eds). Presses de l'Université du Québec, Canada.

Le Breton, D. (1999). *L'adieu au corps*. Métailié, Paris.

Lecomte, R. (2011). Révolution tunisienne et Internet: le rôle des médias sociaux. *L'Année du Maghreb*, VII.

Lespinas, J. *et al.* (2013). Les réseaux sociaux numériques: quel impact sur notre sociabilité? [Online]. Available at: https://journals.openedition.org/rfsic/3495.

Mallein, P. and Toussaint, Y. (1994). L'intégration sociale des TIC: une sociologie des usages. *Technologies de l'Information et de la Société*, 6(4), 315–335.

Malone, T., Laubacher, R. and Dellarocas, C. (2009). Harnessing Crowds: Mapping the Genome of Collective Intelligence. Working document, MIT Sloan School of Management, Cambridge.

Napo, G. (2013). Du processus de démocratisation à la crise de citoyenneté et des institutions de socialisation. *Citoyenneté et transformations sociales en Afrique*, 1, 2(5), 7–29.

Reguer, D. (2009). *Blogs, médias sociaux et politique*. Les 2 Encres, Paris.

Rosanvallon, P. (2006). La contre-démocratie. La politique à l'âge de la défiance. *Le Philosophoire*, 2(27), 257–263.

Tehlwall, M. (2009). Social Network Sites Users and Uses. *Advances in Computers*, 76, 19–73.

Zammar, N. (2012). Réseaux sociaux numériques: Essai de catégorisation et cartographie des controverses. Sociologie. PhD thesis, Rennes 2, Rennes.

## 3.6. Webography

http://news.icilome.com/?idnews=797735 [Accessed March 2018].

http://www.faridanabourema.org [Accessed April 2018].

htp://www.mediassociaux.fr/2009/06/29/une-definitiion-des-medias-sociaux.html [Accessed March 2018].

http://www.relation-transformation-partage.info/wordpress/2010/10/10/essai-de-typologie-des-reseaux-sociaux/ [Accessed June 2018].

http://www.rfi.fr/technologies/20161124-internetreseaux-sociaux-citoyennete-numerique-afrique [Accessed March 2018].

http://www.tutorials-computer-software.com/2009/10/definition-web-10-20-et-30.html.

https://www.cotedivoire.news/benin/20160-lutte-anti-cfa-kemi-seba-annonce-manifestations-afrique-diaspora.html [Accessed July 2018].

https://www.youtube.com/channel/UCZAtrvoju_bYM3fEOes7AEA/videos.

www.courrierinternational.com-videos-cnn-filme-une-vente-aux-encheres-d-etres-humains.

www.koldaews.com/2018/03/08/campage-de-sesibilisation-nittu-degg-lance-sama-baat-a813474.html [Accessed March 2018].

www.lomeit.tg-les-reseaux-sociaux-preferes-au-togo [Accessed June 2018].

www.rfi.fr/afrique/201110228-crise-ivoirienne-vit-facebook-twitter.

www.togoenmarche.com.

www.togosite.com-index-php-togo-8262-togo-pari-reussi-pour-tikpi-atchadam-et-le-pnp.

# 4

# Social Media Challenges in Citizen Participation in Africa

Through its interactive and participatory components such as blogs, discussion forums and microblogs, the Web becomes a "revitalized and expanded public space" (Védel 2003), also called "digital public space" (Maude *et al.* 2017). It is a channel that stimulates debate and facilitates citizen action. The Web allows the public "to edit its own content, decide how to view it and create its own communities of interest" (Augier 2016, p. 126) citing (Vignaux 2008)). Citizen participation is easily illustrated by the possibility, for any source of information, to debate and discuss any subject (Guaaybess 2015). In addition, the digital public space has strengthened the three levers of citizen participation: thinking, acting and debating (Kalugin and Merlant 2017 p. 21) in relation to Rosanvallon's (2006) counter-democratic functions such as monitoring activities (of government action), judgment and prevention. It should also be noted that "the Internet is easily accessible to currents of opinion poorly or not represented in real political systems, which can thus place on the public agenda issues abandoned by parties or the media" (Védel 2003, p. 82).

These theoretical contributions agree on the contributions of social media and the practices and uses of these tools in citizen action to achieve an objective related to a specific event in the context of public space. Such a

---

Chapter written by Oulaï Honoré KAHI.

context underlies our research question: what are the challenges of information and communication technologies, understood in the sense of socio-digital spaces, in citizen participation in African countries?

The objective is to show how socio-digital spaces can provide opportunities for a renewal of public debate and forums for expression. At the same time, we highlight the fact that the new digital public spaces are major assets for social actors in orienting public opinion and intervention on social reality. The aim here is to explore socio-digital systems as legitimate spaces for participation and to identify the multiple challenges and opportunities for Africa. We start from the user-centered reward usage theory (Dandouau *et al.* 2011) to understand the stakes of social media related to the logic of use in the African public space to the point of organizing efficient citizen activities there. Moreover, our reading of the current context of citizen participation in African countries, in relation to the Metcalfe Act (1980), the Reed Act (2000) and Mark Granovetter's (1975) theory of weak and strong ties, leads us to the hypothesis that diverse and real challenges are linked to the use of social media.

We test this hypothesis by using descriptive interpretive research, a qualitative research device for producing descriptions in the context of the phenomena under study and in the search for meaning from the perspective of the people concerned (Gallagher 2010). The results obtained led to the organization of the reflection around the following areas of interest: social media, digital public space, citizen participation, social media issues and digital citizenship.

## 4.1. Theoretical context

We will describe the three fundamental theoretical elements that form the framework of our study: social media, public space and citizen participation.

### 4.1.1. *Social media*

Social media is an integral part of the modes of citizen and political participation. These tools are now part of the social fabric, due to their frequent use and impact.

### 4.1.2. *Overview of the characteristics of the social web and social media*

The advent of social media is the result of the expansion of the social web, an offshoot of the Web 2.0 concept about which Laurent (2008) states: "Web 2.0 is characterized above all by citizen practices, just as it exists only through its users" (p. 208). The social web will also experience an expansion of uses with the rise of social media as its main tools. It is based on the user who creates content, shares it, personalizes it and relays other content. It is at the heart of the communication process, which is the first characteristic of the social web described by Proulx and Millerand (2010). The second is the accessibility of tools and content, the ease of technical appropriation resulting from the functionalities of recent platforms. In addition to these first two, there is a third one: control in favor of the development of user collaboration methods. The last two characteristics are the upheavals that occur in business models in relation to the logic of large numbers, and the multitude of contributions and the plurality of practices of use (prescribed, diverted uses, etc.).

These characteristics are at the heart of the participatory ideology of the social web that found social media. The growth of the latter as the main tools will contribute to the explosion of Web 2.0 uses. These social media are revealed as:

> "[a] set of online platforms creating social interaction between different users around digital content (photos, texts, videos) and according to varying degrees of affinity. They are the focus of all attention, their audience is constantly growing, and they are enjoying an ever-increasing interest from companies and institutions." (Dupin 2010, p. 14)

For Dewing (2013, p. 1), "the term social media refers to a wide range of media Internet and mobile services that allow users to participate in online exchanges, to disseminate content they have created themselves and to join online communities". The concepts and notions used in these definitions all refer to the following five pillars of social media: participation (contribution), openness (collaboration and information exchange), conversation (dialogue and attentive listening), community (interest group) and interconnection (links with other sites, resources or people). These five pillars support the premise that ordinary people can produce and distribute digital content, and justify the terms "media" and "social". The term *media* refers to the fact that electronic channels ensure the distribution and exchange of content, while *social* refers to the linking (permanent flows, interactions) of individuals through socio-digital networks. In this regard, Ruette-Guyot and Leclerc (2009) point out that "these new media break the traditional communication patterns that clearly separate transmitters and receivers. They allow not only reciprocal but also multiple communication: from everyone to everyone" (2009, p. 133). The information is given in a participatory way, reaching citizens through connected objects and allowing new practices. The different types of social media are blogs, social networks, forums, content sharing sites, etc.

## 4.2. *Digital public space*

The public space, an active construction of reality, is continuously falling apart. In its current digital form (Maude *et al.* 2017), it consists mainly of socio-digital media (blogs, discussion forums, etc.) where users can debate and discuss any subject (Guaaybess 2015). In its acceptance of media space, public space is "functionally and normatively inseparable from the role of the media" (Wolton 1992, p. 31) and a symbolic place of mediation between citizens, media and politics. As a result of this mediation which citizens are subject to, Macé (2005) makes the following clarification:

> "The particularity of the mass media is that it is both the stage and one of the actors in the public sphere. It has been the main focus of public debates since it is no longer limited to parliamentary forums and the academic written press, but are now mediated by the mass dissemination of images and media discourses." (2005, p. 126)

Modern public space exists only through the mediation of the press, which operationalizes the principle of advertising – in the Habermasian sense of the term – to citizens. With technological progress, modern digital communication tools or social media make it possible to publicize debates and differing opinions. In this context, a new so-called digital public space, or public cyberspace, imbued with a set of tools, practices, communication and mobilization logics, reconfigures relations between citizens as well as the way of doing and thinking about citizen participation. The massive adoption and use of social media, such as Facebook, YouTube, independent blogs or specialized social networks, are changing the boundaries of the public space and reintroducing citizen participation. Through tools such as blogs, Facebook pages, discussion forums and microblogs, the Web becomes a vehicle for stimulating debate. In these countries where relative freedom of expression is required, forums provide an ideal space for citizen movements (see appendix) to express themselves without hindrance.

Citizens' movements in sub-Saharan Africa such as *Y'en a Marre* in Senegal, *Balai Citoyen* in Burkina Faso, *Filimbi* (whistle in Swahili) and *Lucha* (Fight for Change) in the Democratic Republic of Congo, *Ça doit changer* in Chad, *Ça suffit comme ça* in Gabon and many others are constituted as the African league of web-activists for democracy (Africtivists). They have taken the lead both physically (geographical scope) and digitally (citizen blogs) through media activities in the various digital public spaces.

The citizen platforms developed configure the actions of Internet users by applying action formats to them. In addition, they structure online relationships by promoting interaction. These socio-technical measures define frameworks and provide transparent ways of organizing public action, thus guaranteeing better citizenship.

Thus, with the advent of social media, the public space – a place for discussion in terms of concrete social activities – has undergone many changes and transformations that have led to a variety of tools. This has resulted in institutional forms, mediatized forms and digital forms or digital public space.

## 4.3. Citizen participation

### 4.3.1. *Overview of citizen participation*

Citizen information can enable the community to contribute to the construction of responses and solutions to community needs. These needs will vary depending on the contexts in which citizen participation takes place. Cunningham (1972) identifies three essential dimensions in the definitional approach to citizen participation:

– ordinary people, i.e. members of the community who do not have formal sources of power beyond their numbers;

– a power assumed by these people who lead the group to think and act in the way they think;

– decisions that involve significant and substantive choices which will influence the community.

With regard to these three dimensions, citizen participation is defined by Cunningham as “a process in which ordinary people in a community exercise power over decisions relating to the general affairs of their community” ((1972) in (André *et al.* 2006, p. 2)). For Hardina (2008), it is “a process of involving the most disadvantaged or marginalized people in decisions about the services they receive, that those who represent them take”. This definition emphasizes the beneficial effect of the local control of decisions in response to the effects of economic and social oppression experienced by these groups of people, mainly young people (in African countries), which reflect an acute awareness of their living conditions. The popular demonstrations, social mobilizations and citizen protests of the past decade in sub-Saharan African countries are proof of this. These demands for new governance, transparency and pursual of a common cause show that citizen participation refers more specifically to that of ordinary people, those whose sources of power are limited. It is an “intentional act-together, marked by the clear project of the protagonists to mobilize together” (Nephew 2015, p. 28), specific to social movements or citizen movements that have been highly visible in recent years on social media (see Appendix, section 4.7).

Citizen information on citizen blogs has focused on “promoting people’s ability to participate in the making of decisions that affect them at all levels and in all areas of life in society”. These social media, blogs and Twitter pages, in particular, prioritize the three dimensions:

– stimulate critical thinking;

– contribute to democratic public debate;

– encourage action.

The levels of participation in relation to the guidelines are information, consultation, dialogue, co-production (cooperation and collaboration) and project support. The citizen blogs of the movements of the Africtivist collective specify through the articles and organizational charters the specific logics thus declined: act, mobilize, organize, build, contest, collaborate, popularize, ally, network, preserve, monitor, increase awareness, cooperate, demonstrate, etc. For the most part, these participatory logics are articulated around the visions of the citizen movements of this collective.

### 4.3.2. *Use of citizen blogs*

Citizen blogs are social media that are compatible with citizen participation practices. They constitute new spaces of expression and action for the public, unlike the mass media which impose information. Through these social media, sharing opinions, instant debate and denunciation of illegal practices are possible. This can be seen on blogs affiliated to the Africtivist collective, of which *Balai Citoyen* and *Y'en a marre* are just a few examples.

The technical configuration of citizen blogs is an asset which has contributed to their success in revitalizing citizen participation. Thus, design and structuring lead to their best promotion. On both blogs[1], it is the logos and various images that encourage action or provide information on the context of the fight. The broom, to clean the country of corruption, as well as to literally clean the streets, is the symbol of Balai Citoyen. The Burkinabe flag around the wrist is a sign of territorial anchoring. On Y'en a marre's blog, the raised fist is the symbol of the fed up, to shout indignation, the categorical refusal of bad governance. The Senegalese flag around the wrist also illustrates the territorial anchoring. The navigation thread through the sections in each blog provides access to various information: documents, videos, publications, programs, projects, etc.

## 4.4. Methodology

We use descriptive and interpretive qualitative research. The advantage of such a device is to produce a description of the phenomena encountered, to describe their properties, components and variations in context. The interpretative nature makes it possible to understand the meaning of the phenomena from the point of view of the people concerned (Gallagher 2010). This implies "a concordance between empirical data and their interpretation" (Laperrière 1997) as well as the participants and the means of collection.

1 www.lebalaicitoyen.com and http://yenamarre.sn.

The method used to select participants is the snowball sampling technique, which consists of "making initial contact with key informants who in turn point to information-rich cases" (Pickard 2007, p. 65). These key informants are cyber activists and telecenter managers (internet cafés), considered as cases rich in information. The principle of empirical saturation or data saturation was used for the constitution of this type of sample. In addition, we have used platforms of citizen movements organized in the African League of Web Activists for Democracy (Africtivists).

Data collection was conducted through web-based ethnographic observation (Barats *et al.* 2013), semi-directive interviewing and focus group interviewing (Geoffrion 2009; Touré 2010). Ethnographic observation refers to the use of blogs, discussion forums and Twitter pages. It took place over six months. During this period, we followed the activities of the activists in two-month increments. The semi-directive interview was conducted with cyber activists. The aim is to achieve a thorough understanding of their publishing practices. The main principle of the focus group interview is an interaction between participants as a means of collecting (Baribeau and Germain 2010) to document opinions on citizen participation. Thus, we organized meetings with five small groups (four people per group) with whom discussions took place. Interactions between participants made it possible to collect opinions (excerpts from conversations) on citizen participation practices.

The data analysis approach adopted is a qualitative content analysis (Mucchielli 2006; Bardin 2007). This aspect of the study is mainly based on the content analysis of the digital traces and conversation extracts of the respondents surveyed. We use, among other things, the following units of meaning at play: type of socio-digital spaces in use, forms of participation, repertoires of collective actions and various issues. The aim is to identify the links between these units of meaning, to explain them and to account for their meaning.

The data collected highlight the four areas of practical challenges around which social media contributions are articulated.

## 4.5. Main practical challenges

These challenges are communicative, political, social and technical. We present them in the following sections.

### 4.5.1. *Communication challenges*

The framework of the challenges is based on the contributions of social media in terms of communication and interactivity. In other words, the linearity that has characterized the functionalist approach to communication is giving way to the interactive paradigm. The new way of communicating with social media requires us to "go beyond the diffusionist approach to focus on listening, analysis, appropriation and mutual influence" (Maisonneuve 2010, p. 20). As a result, social media involves individuals in isolation or in small groups or in citizen movements, in a new communicative logic: from several to several. Respondents argue that digital tools such as Facebook, Twitter and forums allow them to reach multiple people at once and facilitate the delivery of event announcements in a short period of time.

This "all to many" makes the receivers transmitters and the informed informers (Walry 2014). There is a flow of information and communication exchanges that make it easier to carry out various processes: continuous information attitudes, consultation (gathering the opinions of the audiences concerned), consultation (exchange of points of view) and co-production (citizen projects). These activities are made possible thanks to the multiple functionalities of social media. This is what respondents explain in their use of these media: "On Facebook, we find techniques to put our comments and posts online easily."[2]

The Facebook pages of citizen movements show multiple functions that are used in different ways: "About", "Photos", "Events", "LiveChat with us", "Publications", "Articles" and "Communities".

2 The interview verbatim was deferred without modifying the style of oral conversation and language terms specific to their lexical universe.

These Facebook page functions alone are sufficient to adapt methods of citizen participation in the web environment. This will include collecting information from citizens, receiving feedback on needs for improvement and using innovative ideas. There is also a particularity in the types of comments on Facebook that promote interaction:

– comment attached to the profile (text box: what's on your mind);

– comment on another comment or video (three short sections: "like", "comment", "share"). By clicking on "comment", a text box appears next to the profile where it reads: "write a comment";

– a comment that follows a video or a posted article. A window appears with the words "write something".

These functional characteristics highlight new modes of exchange between users on the Facebook pages of citizen movements.

Twitter is used for short messages as alerts or announcements, event reminders or mobilization. Small individuals or groups use microblogging in citizen participation to share messages with each other. Citizen movements have very up-to-date Twitter pages; therefore very active. These pages and Facebook pages contain hypertext links (often embedded in messages) to videos, articles and other types of publications.

Hypertext links are numerous in both tweets and posts on Facebook pages and do not link to the same target sites. These messages (containing the links) are therefore open exchanges; this intensifies interactions with a wide audience on various topics and generates traffic. One respondent stated: "Links on Facebook and especially Twitter are very good and useful for us; there are many videos, links and photos." These comments reflect the importance of using social media in the sharing of content and its distribution and justify the strategies for using such tools. In addition, there is strong participation on both types of social media for only the five citizen movements (as of June 25, 2018).

| | Facebook | | Twitter | |
|---|---|---|---|---|
| | **Likes** | **Subscribers** | **Likes** | **Subscribers** |
| **Lucha** | 79,309 | 80,393 | 6,682 | 117,000 |
| **Y'en a marre** | 61,717 | 62,513 | 479 | 9,216 |
| **Balai Citoyen** | 52,684 | 52,153 | 510 | 7,376 |
| **Filimbi** | 43,210 | 43,520 | N/A | 25 |
| **Ras-le-bol** | 4,457 | 4,487 | N/A | N/A |

**Table 4.1.** *Evaluation of traffic in terms of "likes" on the selected spaces (source: Facebook pages and includes Twitter from the citizen movements studied)*

These data highlight the reality of the communication challenges related to social media according to the laws of Metcalfe (2000) and Reed (1980). Robert Metcalfe's law expresses the added value of networks: it increases quadratically with the number of people connected and linked. It is the number of links, the number of possible relationships that matters. This law works well to explain the concept of the one-to-one relationship and not that of one to many, let alone "several to many" because it applies both within communities and between communities themselves (Huet 2009, p. 108; Digout and Besson 2016).

This is what prompted David Reed to formulate complementary legislation. According to Reed, networks that encourage the construction of communicating groups create value that grows exponentially with the size of the network. This law adapted to networks or individuals, citizen movements and communities is considered. Thus, it takes into account social media in their interactive and participatory components.

Even if Facebook and Twitter are not used in the same way, it must be recognized that technologies have made a considerable contribution to the development of communication strategies focused on citizen participation. In terms of communication challenges, this online participation intends to use the means at its disposal (citizen blogs, Facebook pages, Twitter, etc.) and propose content and contribute to the groups' activities.

### 4.5.2. *Political challenges*

We now show how social media can achieve a "real public space" (advantage) with citizen participation by examining negative effects (risks) for political purposes.

#### 4.5.2.1. *Reaching the real public space*

Political branding will push national media public spaces and the relative freedom of expression will push many Africans to Facebook (an essential part of their daily lives) to express themselves and stay informed.

Some of the characteristics of this digital platform facilitate political communication and bring together a large number of people despite geographical, social, cultural and other distances, etc. These multiple functionalities make Facebook a field of possibility in the digital public space. A study by USEO (2011) classified the uses offered by social media into the following three categories:

– influence uses: the objective is to make the members' activity visible and enhance their activity, to enhance the most interesting content and to ensure the dissemination of information according to natural channels;

– collaboration uses: these uses aim to bring members into contact, set up and operate interest groups and provide spaces for exchange;

– coordination uses: they make it possible to organize events, manage the production of materials and facilitate field actions in the broadest sense.

These uses allow both politicians and citizen movements to carry out the following activities in the public space:

– stimulate reflection on subjects chosen by citizens;

– discuss community issues with representatives of various communities and citizen movements;

– get reactions from social groups on the words and decisions of politicians;

– build links between stakeholders and community interests.

It is therefore possible to engage in a flexible way, through social media, in an interactive dialogue with citizens in general and constituents in particular. This is explained by the cyber activists in our study sample. These activists and other respondents who have a Facebook account indicated that the following features are the most commonly used: publishing content, writing content directly into the interface, commenting or sharing information and videos, being responsive on the walls of friends or political representatives, etc. Almost all respondents focus on publishing content: news sequences, announcements of mobilizing citizen activities, political topics, etc.; users also comment on published posts and follow links to other sites, blogs or microblogs.

Facebook's flexible use of content publishing provides any individual with a communication tool that has immediate effect in the public space and is therefore ideal in response to political events and news. Facebook is often used in African countries – outside of personal needs – to raise awareness of particular causes and facilitate public scrutiny of political actors and institutions. For example, in Côte d'Ivoire, the increase in electricity consumption billing has led to widespread anger. Various opinions have been expressed on Facebook to mark overall dissatisfaction. Koffi *et al.* (2018, p. 423) emphasize this point:

> "[...] they multiply mobilization actions by calling for national disobedience for non-payment of the electricity bill. Their show of strength on Facebook was so resounding that it brought consumers together in the physical public space to demonstrate. These generally uncontrolled mass movements have degenerated into breakdowns and looting of the Ivorian Electricity Company's branches as well as other companies and banks located nearby..."

It is an action of surveillance – citizen vigilance, denunciation of abuse by political authorities, evaluation of their actions (Rosanvallon 2006, p. 75) – in the public space that was followed by that of judgment. This judging function has been illustrated by the permanent holding of a real court of opinion (Rosanvallon 2006) through publication practices as pointed out by Koffi *et al.* (2018):

> "Reactions on the walls reflect their dismay: 'Why increase electricity? While we sell electricity to neighboring countries and they have it cheaper.' Another Internet user, Myriam, says the following: 'No, it doesn't work like that. When a commodity is scarce, it is expensive, and that makes sense

> because it is difficult to find, but if it is plentiful, the price drops because it is available everywhere. It should be cheaper for us.'" (2018, p. 422)

This has resulted in the application of the activity prevention function, which aims to render a political decision null and void, to avoid or circumvent the rule (Rosanvallon 2006). Koffi *et al.* (2018, p. 423) add:

> "In the face of this social pressure, the President of the Republic canceled the increase measure in his speech of May 1' 2016: 'The increase in electricity in January is a problem… I have decided to cancel this increase outright and I apologize for this injustice.'"

Facebook has thus made possible forms of citizen participation in the digital public space by being a "transmission belt for rational debates, making it possible to publicize them, to include other citizens, as well as to transmit this public opinion to the authorities in place" (Habermas 1997, pp. 191–192). There is an interactive dialogue with users, with communication being at the center of the debate.

The traditional media used in politics seem to be giving way to the social media that impact them, ranging from complementarity to serious competition. In Africa, due to the accessibility of objects connected to Wi-Fi and the Internet offers brought by modest grants, the scoop is reserved for web media. Users connect to Facebook in the early hours of the day. Social media is thus becoming an essential alternative (influential media).

Although these media can make it possible to build citizen movements evolving outside any state control, it must be noted that there are real risks in the use of these digital tools. We mention, among others: blocking access to Facebook, political censorship, computer warfare techniques, distortion of information, etc.

### 4.5.3. *Social challenges*

The challenges identified here refer to the relevant contributions resulting from the use of social media: friendship, social capital, strength of ties and other rewards.

Africans generally have a social group culture, ferment for a legendary sociability. The arrival of Facebook is a catalyst for online social relationships, understood in the sense of interaction, communication and socialization. Almost all respondents say they are on Facebook to experience public relations, a form of sociability that focuses on friendly networks. They use Facebook "because it's a space of contacts, of friendly ties", according to one student interviewed. "We have a degree of freedom and choice of friends," she continues. "This freedom is not acquired; there are people who refuse friendship requests," says another, who states: "The network often sends out make sure you know these people before you send them friendship requests." This is the problem of homophily. This concept is as follows:

> "The tendency for friendships to form between individuals who look alike on a specific aspect. It reflects the social inscription of friendship; homophily rates reflect an overall tendency for friendships to conform to a norm in terms of social preference." (Spencer 1993 in Digout and Besson 2016, p. 42)

The affinities in this sense are very varied, even if Facebook tries to offer links on the basis of a social computation understood as the construction and sharing of collective digital memory – photos, videos, music – (Coutant and Stinger 2009). These sharing activities are most often aimed at building social capital. Social capital refers to "all the resources inserted in the social networks available and used by the actors for action" (Fuhrer 2011). It is the structural dimension (links between member actors) that shapes the uses on Facebook. Thus, having friends allows you to gain prestige. Weaving and maintaining friendly and loving relationships is a guarantee of maturity and autonomy. The likes received express the feelings of the peers. The number of likes reflects a quantity of credits and attention that can be capitalized or added. Online mediatization and self-advertising are a strategy for enriching social capital (Foessel 2008). This social capital is not necessarily correlated to relational dimensions (trust and shared norms) which, in the end, are neither efficient nor operational. This predominance of the structural dimension of social capital explains the existence of the "strength of weak ties" (Granovetter 1975) that characterize Facebook relationships. For this author:

> "A network is made up of strong and weak ties. Strong ties are those that we have, for example, with close friends (this is the sustained and frequent relationship). Weak ties are mere acquaintances. Weak ties are said to be strong insofar as, if they

> are diversified, they make it possible to penetrate other social networks than those constituted by strong ties." (Granovetter quoted by Digout and Besson 2016, p. 52)

Thus, if "friends" on Facebook are in strong ties – family ties and close friendships – they are likely to become friends in the "real world" if they meet each other, which often happens, but not always. Moreover, one of the strengths of social media in Africa is to allow "social chat", which is part of the oral culture: publishing a post that is not relevant, but is interesting because it encourages a large number of people to react. It is an incentive to chat that often leads to heated comments, creates disconnection anxiety and alone justifies frequent use of Facebook.

Regarding the reasons for use, interview excerpts showed that, for most Internet users, using social media is explained by the particular benefits they derive from it. These benefits are consistent with the user-centered "reward usage" theory (Blumler and Katz 1974; Dandouau *et al.* 2001). These authors have identified four categories of social media uses and rewards: personal identification (personal references, value reinforcement), social relationships (communication, social interaction, affection and socialization), distraction/entertainment and monitoring (Therrien 2012). These orientations explain the increase in Facebook's usage rate, the degree of connection to Internet networks and the constant use of social media. Social challenges underlie the logic of using social media.

### 4.5.4. *Technical challenges*

We mentioned above the characteristics of the social web as described by Proulx and Millerand (2010). The first two underlie the technical challenges: content creation, personalization, sharing and ease of technical appropriation of functionalities. Citizen blogs, forum interfaces and Facebook pages have "helped to shape new social mobilization practices" (Proulx and Millerand 2010, p. 158). An examination of the platforms built by citizen movements shows that they have used production and dissemination opportunities to bypass traditional media. The digital communication tools developed made it possible to deploy their repertoire of collective action, which was diversely formulated around the following tactics: recruitment of members, mobilization of members for collective action, development of effective communication and tactics strategies.

We consider the four platforms of the *Y'en a marre* citizens' movements, Balai Citoyen, Ras-le-bol Congo and Lucha to be examples. We analyse three dimensions: attractiveness, need and ownership. The attractive dimension refers to the interests and expectations of users. These include logos, colors, headings (layout and title), legibility and ease of navigation. In addition, the availability and relevance of the expected information are other elements that make up the attractive dimension. Moreover, these blogs constitute the satisfaction of information needs that can be described as "psychological and social needs of uses" (Jouët 1992, p. 27). These needs justify the appropriation of the use of these blogs in a trial and error-free and intuitive way to acquire know-how. According to their comments, the users of these citizen spaces have no difficulty in handling them.

We can therefore agree with Badouard (2012) that digital technologies of participation, in this case citizen blogs, "take action" in the sense that they configure and guide online actions. Internet users are thus led to act on published content and "interact" with each other. The interfaces organize social relationships between the categories of actors through online exchanges. The logics of the activities declined in the action repertoires underlie the forms of engagement in the public space.

## 4.6. Conclusion

We have analyzed the way in which social actors invest in new public spaces to guide public opinion and act on social reality. An exploration of socio-digital devices as legitimate spaces for participation and a determination of multiple challenges and opportunities for Africa were established. The interpretive descriptive research methodology – a qualitative research device for the production of descriptions in the context of the phenomena under study and in the search for meaning – has made it possible to obtain field results. Thus, a description of social media, digital public space and citizen participation was presented. Communication, political, social and technical challenges related to citizen participation were the subject of a contextual development based on field data. The user-centered reward usage theories, Metcalfe's law, Reed's law and Mark Granovetter's theory of weak and strong ties have been used to understand social media challenges related to the logic of use in the African public space for the purpose of organizing efficient citizen activities.

## 4.7. Appendix

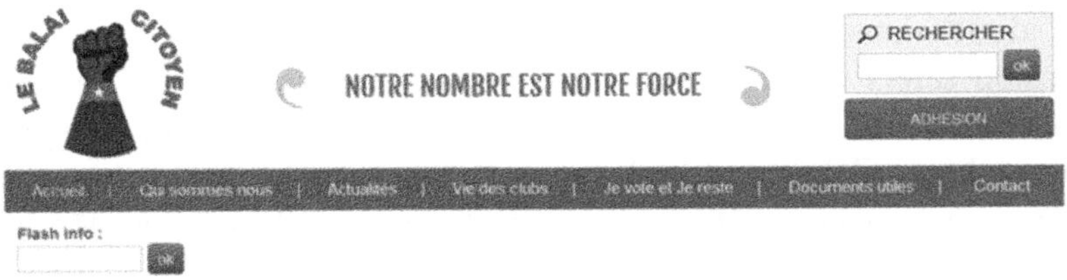


**Figure 4.1.** *Blog of the Balai Citoyen citizen movement in* Burkina Faso[3]: Notre nombre est notre force – *Our number is our strength. For a color version of the figures in this book, see www.iste.co.uk/atchoua/digital1.zip*

**Figure 4.2.** *Blog of the* Y'en a marre *citizen movement in Senegal*[4]*:* Il n'y a pas de destin forclos, il n'y a que des resonsabilités désertées – *There are no precluded destinies, only deserted responsibilities. For a color version of the figures in this book, see www.iste.co.uk/atchoua/digital1.zip*

**Figure 4.3.** *Blog of the Lucha citizen movement, in DR Congo*[5]*:* Lutte pour le changement – *Struggle for change. For a color version of the figures in this book, see www.iste.co.uk/atchoua/digital1.zip*

3 www.lebalaicitoyen.com.

4 http://yenamarre.sn.

5 www.luchacongo.org.

**Figure 4.4.** *Blog of the Filimbi citizen movement, in DR Congo*[6]. *For a color version of the figures in this book, see www.iste.co.uk/atchoua/digital1.zip*

**Figure 4.5.** *Facebook page of the Lyina citizen movement. For a color version of the figures in this book, see www.iste.co.uk/atchoua/digital1.zip*

**Figure 4.6.** *Facebook page of the Ras-le-bol citizen movement. For a color version of the figures in this book, see www.iste.co.uk/atchoua/digital1.zip*

6 https://www.frontlinedefenders.org/en/profile/filimbi.

## 4.8. References

André, P. *et al.* (2006). Participation citoyenne [Online]. *Le dictionnaire encyclopédique de l'administration publique*. Available at: www.dictionnaire.enp.ca.

Augier, M. (2016). *La société numérique*. L'Harmattan, Paris.

Bonenfant, M. *et al.* (2017). *Les pratiques transformatrices des espaces socionumériques*. Presses de l'Université du Québec, Quebec.

Digout, J. and Besson, L. (2016). *Web social – Le Web 2.0 au service de la création de valeur*. Vuibert, Paris.

Dupin, A. and Epelboin, F. (2010). *Communiquer sur les réseaux sociaux. Les méthodes et les outils indispensables pour vos stratégies de communication sur les médias sociaux*. FYP, Paris.

François, L. (2008). *Marketing 2.0: l'intelligence collective*. M21, Paris.

Granovetter, M. (1975). The Strength of Weak Ties. *The American Journal of Sociology*, 78(6), 1360–1380.

Guaaybess, T. (2015). *La relation dynamique entre technologie est-elle évidente? Le cas des révolutions – dites Web 2.0 – pour la démocratie dans les pays arabes*. L'Harmattan, Paris.

Habermas, J. (1997). *L'espace public : archéologie de la publicité comme dimension constitutive de la société bourgeoise*. Payot, Paris.

Huet, J.-M. (2009). Quel avenir pour les réseaux dans le management?. *L'Expansion Management Review*, 133, 104–113.

Kalouguine, T. and Merlant, P. (2017). Informer… sans être journaliste. Guide pratique et éthique en 50 questions [Online]. *RITIMO*. Available at: https://www.ritimo.org/Informer-sans-etre-journaliste.

Koffi, H.B.D.I., *et al.* (2018). Facebook en Côte d'Ivoire: nouvel espace d'expression et de pression des citoyens sur le gouvernement? [Online]. *Réseaux sociaux, traces numériques, communication électronique*. Available at: https://eviolence.hypotheses.org/files/2018/06/Sommaire_RSTNCE2018.pdf.

Leclerc, S. and Ruette-Guyot, E. (2009). *Web 2.0: la communication "Iters-@ctiv"*. Economica, Paris.

Macé, É. and Maigret, É. (2006). *Penser les médias cultures. Nouvelles pratiques et nouvelles approches de la représentation du monde*. Armand Colin, Paris.

Maude, B. *et al.* (2017). *Les pratiques transformatrices des espaces socio-numériques*. PUQ, Montreal.

Neveu, E. (2015). *Sociologie des mouvements sociaux*. La Découverte, Paris.

Poulx, S., Milette, M. and Lorna, H. (2012). *Médias sociaux, enjeux pour la communication*. Presses de l'Université du Québec, Quebec.

Proulx, S., Millerand, F. and Rueff, J. (2010). *Web social, mutation de la communication*. Presses de l'Université du Québec, Quebec.

Rosanvallon, P. (2006). *La contre-démocratie: la politique à l'âge de la défiance*. Le Seuil, Paris.

Useo (2011). Les réseaux sociaux des mouvements politiques au cœur d'une acculturation de nos démocraties modernes [Online]. Available at: http://www.lecko.fr_etude-reseaux-sociaux-des-mouvements-politiques.html.

Vignaux, G. (2008). Web 2.0 et les blogs de savoirs [Online]. Available at: https://www.academia.edu/2972535/Le_Web_2.0_et_les_blogs_de_savoir.

Védel, B. (2003). La citoyenneté et ses conditions d'expression, l'internet et ses usages citoyens. *Cahiers français*, 316.

Wolton, D. (2008). Les contradictions de l'espace public médiatisé. *L'espace public*, CNRS, Paris, 31–58.

# 5

# Medialism in the Societal Space of John Updike

Medialism is "a form of writing that fuses the media [where] strictly literary material and media materials [combine] together in the literary nature of the text" (Guiyoba cited by (Dého 2014, pp. 186–187)). Thus, as a result of new concepts of "globalization", we are reading American literature that is increasingly dictated by information and communication technologies (ICTs). Kathryn Van Spanckeren rightly said that "American literature has come a long and winding way from the pre-colonial era to the contemporary era. Society, history, technology have all had a profound influence on it" (Van Spanckeren 1994, p. 166). This massive invasion of the literary world by the media or "strong literary intermediation" (Guiyoba 2015, p. 7) generates a certain fraternization within the media. John Updike's work is presented as a melting pot of sociability, communication and mediation between different protagonists who freely express themselves on the current events of the facts that interest them. What is the point of such a writing practice? How does the media scripting in John Updike's novels help to conquer the public space through reading? What is the impact of these media on the reader and thus on society in general? Finally, what makes Updikian stories unique?

This contribution aims, first, to identify the relevant media and their effects on the narrative, and ultimately to present the generic formal innovations that devote the novelistic universe to a cultural transfer in a globalized world. It formulates the hypothesis that the intrusion of the media

Chapter written by Olivier Koffi KOFFI.

into the textual fabric contributes to the concern for renewal and originality of fiction writing.

In an attempt to provide answers to these questions, our analysis was based on a qualitative method and transcultural poetics initiated by Josias Semujanga. For content analysis, it should be noted that the study has a corpus composed of three novels by John Updike, namely *Rabbit, Run*, *Rabbit Redux* and *Rabbit is Rich*. On the one hand, this allows us to present a chorus of media voices and, on the other hand, offer media scriptures of fragmentation in a social and cultural mediation. These three novels offer the reader an exaggerated narrative that shows several types of media with a writing style of repetition, variation and even an art of wandering. John Updike's success is largely based on these novels. The most prestigious literary prize in the United States, called Pulitzer, awarded to the author for *Rabbit is Rich,* is a good illustration of this. As the qualifier suggests, the qualitative method aims to analyze the content of Updikian texts. The goal here is to grasp the meaning of the media discourse and to understand the meaning of the dissemination of information through the narrative after its interpretation and appreciation. With this method, we are also interested in the values, intentions, goals, objectives and ideologies of the characters in John Updike's novelistic society. It is probably from this perspective that Jean-Pierre Deslauriers writes:

> "[The qualitative method] focuses instead on the analysis of social processes, on the meanings that individuals and communities give to action, on daily life, on the construction of social reality." (Deslauriers 1991, p. 6)

## 5.1. A chorus of media voices

From *Rabbit, Run* to *Rabbit is Rich* to *Rabbit Redux,* the reader is confronted with a receptacle of telephone, television and radio discourses with a mixture of all kinds of newspaper articles. The Updikian hero Harry Angstrom who is a young man constantly listens to the news of his country from his car during his various trips. In these novels, books are represented in a printed or print magazine form. Magazines include the *Consumer Report* or *Bulletin du consommateur*. Different cultures meet and overlap. The narrative journey exposes the illusion of the first mission of Apollo 11 and the landing of the first man on the Moon live on television. The media present a typical business environment, and meetings are planned between salespeople and customers. The literary landscape offers the reader a narrative polyphony. It is in the poetics of Dostoyevsky by the Russian philosopher and theorist Mikhail Bakhtin that the word "polyphony" first appears. According to him, this word:

> "is traced from the Greek *poliphônia* and means according to the etymology 'multiplicity of voices or sounds' at work in a literary text. Thus, it is fair to say that polyphony is necessarily dialogical. It can only be complete with the conjunction of coordination 'and' which associates other entities such as: polyphony and dialogue, polyphony and enunciation, polyphony and intertextuality, polyphony and literary genres." (Bakhtine 1970)

In John Updike's novels, polyphony creates a chorus of narrative media voices. These voices allow the resonances of reality to amplify and nourish each other. The story becomes a puzzle, composition and variation. The process of fiction creation is being swept away by several media voices. Here, "Intermediality takes advantage of the benefits offered by technological progress. It is a piece of writing that integrates and exploits the mass media and various modern communication media." (Atcha 2011, p. 55)

The intrusion of the media into the structure of Updikian texts makes it possible to read literary intermediality. This is what Rajewsky describes as the *as if character* (Rajewsky 2005, pp. 43–64) or as an intermediate reference: it is the idea of finding the deep structure or characteristic features of media in a literary text. The author creates the illusion of the media while playing on a certain mediation between the different characters in middle-class America. In these novels, the media are channels of technical and digital communication.

Thus, "through the product speech, the reader has the impression of reading a newspaper, listening to the radio [or] a telephone conversation [or] watching television" (Atcha 2009, p. 52). The intermediate reference is essentially concerned with the illusion of the media and media configurations in the novel. Intermediality is thus taken in its restricted sense, because "a narrative materialized in a piece of writing contaminated by the information media and by cinema [is] an intrinsically intermedial narrative" (Mariniello 2003).

### 5.1.1. *The voice of the radio: give-and-take*

Radio is a medium that overloads the structures of Updikian texts. A radio receiver is a device that receives radio waves where modulation can be extracted and sounds can be reproduced on a loudspeaker. As defined in this way, radio offers "give-and-take" on several topics in John Updike's novels: politics, music and entertainment. From then on, we witness syntactic and semantic interferences that refer to sound voices in the literary world, as the following excerpt indicates:

> The music on the radio is soothing now, lyrical and unadvertised, and, coming first from Harrisburg and then from Philadelphia, makes a beam he infallibly flies in on. […]
>
> Throughout the early morning, those little hours that are so black, the music keeps coming and the signs keep pointing. His brain feels like a frail but alert invalid with messengers bringing down long corridors all this music and geographical news. (*Rabbit, Run*, pp. 37–38)

Harry Angstrom can be described as a "radiophile". He is constantly guided and maintained by the radio in its various walks. Here, the novel is to the rhythm of lyrical music. Thus, radio gives a certain tone to the literary text that breaks the linearity of the narrative. The storytelling of this song makes it possible to neutralize the authenticity or purity of the work in order to naturalize its metamorphosis. Radio overwhelms textual structures and creates a landscape with sensational and emotional effects. Under the influence of noises and pleasures of all kinds, the novel is carried away by a certain musicality.

### 5.1.2. *The voice of the telephone: a telephone narrative*

The telephone is the other voice that gives a special character to Updikian writing. Here, the story is not only heard, but also communicated. Indeed, the telephone is a communication device designed to transmit the human voice and allow remote conversation. It is a tool that requires a terrestrial or spatial infrastructure and a network for its operation. "Communication [requires this]: sender-message-receiver." (Méchoulan 2003, p. 17)

In John Updike's novel, the telephone often replaces the narrative with a mixture of codes. The progression of the story is based on telephone conversations, and the interpretation of the voice is not left to the reader's skill. The telephone dictates a certain dictatorial power to the diegesis and a certain rhythm is born within the textual fabric favored by the sounds of different communications. The author ensures the autonomy of his texts with a number of varied elements necessary for the internal understanding of the work.

Intradiegic voices give a certain rhythm to the narrative and sometimes interrupt the reading. The authenticity of the novel is thus challenged by the submergence of a multiplicity of sounds. The writing loses its purity under the interweaving of various borrowings from the audio world. For example, in his narrative journey, the reader can witness this telephone excerpt between Pastor Jack Eccles and Harry Angstrom that announces sad news (see the next excerpt):

> "Hello"
> "Hey, Jack? This is Harry Angstrom. I hope I'm not interrupting anything."
> "No, you're not."
> "You don't have any old ladies sitting around sewing or anything, do you?"
> "No."
> "Why, I've been trying to call my apartment and nobody answers and I'm kind of nervous about it. I didn't spend last night there and I'm getting sort of a prickly feeling. I want to go home but I want to know if Janice has done anything like call the cops or anything. Do you know?"
> "Harry, where are you?"
> "Oh, at some drugstore in Brewer." […]
> "I guess to save time I'd better tell you over the phone," he begins. "Harry. A terrible thing has happened to us." (*Rabbit, Run*, pp. 268–269)

Harry Angstrom informs his friend Jack Eccles that he has not been home since last night. He's worried about what Janice will do to him when she gets back. Eccles took advantage of this phone call to tell him some sad news. Harry does not react to what Jack says, but he just listens to the words as they are said. This communication features two protagonists whose voices animate this part of the narrative. The story is thus dialogued, instead of being narrated through a medium. Thus, according to Müller, "the game of being in between" (Müller 2010, p. 18) is born in the narrative.

All the words and expressions selected in this telephone conversation are clues used to confirm this multiplicity of audible elements in the semantic field. The mimesis is a realistic and selective perception in the sense that it reveals a frame of reference within which the "game of telephone exchanges" is used to build diegesis. The way the story is organized creates the effect of reality. The introduction of the foreign element, namely the telephone medium, is an integral part of the writing.

We are witnessing an intermediate text where a thought is already elaborated and writing takes shape. The novel text sweats under the effects of communication noises that sometimes plunge the reader into the characters' language games. Thus, the recurrence of the passages of dialogue tends to transform the narrator into a simple "director" and the reader into a spectator. The narrative of the telephone voice is both a reading of the foreign culture and a writing of this one. *Rabbit, Run* is a receptacle for this type of conversation. Pastor Jack Eccles and Harry are constantly talking on the phone about Janice's birth. The Updikian romantic landscape thus guides the reader towards social facts.

Long telephone excerpts constitute a sound voice in John Updike's fiction space and report on Janice's delivery in a Brewer hospital. Harry and his friend Eccles arranged to visit the new-born. The novel thus refrains itself from a closed author's monologue to give rise to a dialogue in which several voices intertwine, respond to each other, confront each other and are staged by the author. Here, the narrator not only creates the illusion of the characters' voices in the reader, but also tries to create the illusion that there is verbal interaction between them. We observe the issue of restricting narrative information in order to favor the communicative authority. The story is thus telephonic instead of a classic narrative.

In total, radio offers "listening" in the story, when the phone presents a dialogic or telephonic narrative to expose the textual landscape to a multiplicity of voices. Among these voices, the audiovisual sector is part of a dynamic of polyphonic authority.

### 5.1.3. *The voice of television: literature on screen*

Television is a medium that is strongly present in the Updikian narrative. With it, the reader experiences the literature on screen. Television provides the novel with diverse programs: political news, films, advertising sequences, etc. The transmission to the reader of these visual programs and reports, with the associated sounds, makes it possible to speak with an audiovisual voice in John Updike's fiction creation.

Here, the narrator uses the power of the image to establish a certain truth within the story. The use of visual scenes helps the word to put into narrative what is representative. The television voice breaks or cracks the novel from the inside. The linearity of the narrative is interrupted, and the reader's mind wavers between a literary landscape and the audiovisual world. This swing from one side to the other in reading disturbs concentration. A typology of borrowing is being established that requires an audiovisual *background* for understanding. In this case, "intermedial practice introduces a new order of discourse: the disorder of heterogeneity and fragmentation" (Atcha 2009, p. 60).

Television, so to speak, "speaks several languages" in the novel. These different voices create a certain animation, a cadence with sensational and emotional effects in the textual space. The process of constructing the structures of the novel competes with sounds and noises of all kinds to create a kind of symbiosis and harmony in the narrative. We then witness the vision scenes that Updike follows through with several descriptions that give the impression of actually watching television. This "laughter and applause" program is an example, as the following excerpt states:

> So for thirty minutes he sits by her bed watching some crew-cut M.C. tease a lot of elderly women from Akron, Ohio, and Oakland, California. The idea is all these women have tragedies they tell about and then get money according to how much applause there is, but by the time the M.C. gets done delivering commercials and kidding them about their grandchildren and their girlish hairdos there isn't much room for tragedy left. [...]
>
> It isn't too bad a show; a pair of peroxide twins with twitchy tails push the women around to various microphones and booths

> and applause areas. It even makes for kind of peace; he and Janice hold hands. (*Rabbit, Run*, p. 217)

Television sets a tone of entertainment and humor in this excerpt through this game that gives women the opportunity to be rewarded by telling their tragedies. The story remains spectacular and creates a resonance effect. The discourse elaboration process determines the entire nature of the transmission of the audiovisual effect and leads to impressions on the characters.

Through this program, we can already see that Janice and Harry are holding hands as a sign of peace. The television voice thus creates an atmosphere in the novel. In addition to these representative voices in the narrative process, others such as music, film, multimedia, magazines and newspaper articles that mark John Updike's fiction creation can also be mentioned. All these voices that reflect society participate in the development of polyphony within the textual fabric. These poetics of the fragmentary text leads to hybridity, which gives a particular esthetic to the work.

## 5.2. A media description of fragmentation in social mediation

According to Pierre Garrigues, the word "fragment" is linked to the idea of "fragmentation suffered" (Garrigues 1995, p. 409). Thus, the origin of the fragment as part of a whole leads us to think of it in tension with the symbolic unity of the text. Indeed, any fragment sensitively summons the reader into the interaction it produces during its reading. The strategy of the textual object is visually captured in the Updikian narrative as indicated in the television excerpt from the following excerpt:

> Now she sits all day watching the afternoon soaps with Mom-mom and sometimes Mom, *Search for Tomorrow* on Channel 10 and then *Days of Our Lives* on 3 and back to 10 for *As the World Turns* and over to 6 for *One Life to Live* and then 10 again for *The Guiding Light*, Nelson knows the routine from all those days before they let him work at the lot. (*Rabbit is Rich*, p. 295)

This fragment allows the reduction of the structure of the text, which allows the information on the different television channels to be played back in a loop. The route of various programs from channel 10 to channel 10, including channels 3 and 6, introduces a certain segmentation within the story. In this way, multiple voices are evoked in the novel that makes it possible to create cadence and rhythm in the literary text.

John Updike's fragmentation is also this shift in temporal fluency and the distancing in which the narrator who speaks becomes more pluralistic. Here, self-awareness only occurs through the return to oneself from the other and through which the latter becomes one's own disappearance. Here, we find a relational movement between the singular and the plural that fragmented mobility allows. The narrator plays with epanorthosis or metanarratives to allow the reader to experience neutral attributes.

Neutralization requires the neutrality of the television fragment, which, through the disconnection of the cut and interruptions, accentuates the plurality of points of view in the textual landscape. It then immerses in the narrative an ultimate meaning in the descriptive flashes of real effects. The fragmentary enunciation is explained in the narrative interludes of the story. The existence of the fragment shows the communicative value of the form as a possibility of a word in the expressiveness of the visual.

In addition, the word ordering is due to the stabilization of the segmentation of the visual units. The distinction between the arrangement of the novel and the fragment calls into question this apparent unity in the reading space. The use of typographical elements contributes to the construction of the text. As indicated in the excerpt above (*Rabbit is Rich*, p. 295), it is the "italic" that has been used in the various program titles and which characterizes this reading gesture, which is less concerned with the joining of the dispersed elements than with opening up the dispersion of the text.

The fragmentary writing in the Updikian narrative is part of a depiction of the discourse in its visual form. To attach itself to the fragment's method of being in its spatial development is to question the phenomenon in a context where what is to be said and what is said to overlap. The literary fragment shows its intentions without developing them. It rejects any *doxa* but, at the same time, it points out the danger of creating one from scratch, because by wanting to say very little, often what is said often becomes a word of order under the media coverage.

In addition, the heterogeneous Updikian narrative depends on the use of different insertions. John Updike uses several embeddings: the introduction of the telephone, television, music, cinema, multimedia, newspapers and magazines helps to confuse the reader and highlight the heterogeneous and hybrid aspect of the novel. These snippets of history are present one after the other and in alternation with various narrative sequences. These novels contain erotic scenes, or proverbs and theoretical thoughts.

These novels are far from being united texts. They are disturbed or confused by the intermittences of parasitic elements that give the impression of a long series of collages made without any concern for the logic of the narrative. This insertion process disrupts performance or narrative balance. All this leads to fragmented, divided and dislocated texts. For John Updike, there is a density and richness in the layout and text layout of novels. Rabbit's various walks allow him to live with the radio in his car, as shown in the following excerpt:

> Still, his Corona is a snug machine. Solid big-car feel, padded tilt steering wheel, lumbar support lever for adjustable driver comfort, factory-installed AM/FM/MPX four-speaker radio. The radio is what he enjoys, gliding through Brewer with the windows up and locked and the power-boosted ventilation flowing through and the four corners of the car dinging out disco music as from the four corners of the mind's ballroom. Peppy and gentle, the music reminds Rabbit of the music played on radios when he was in high school. […] (*Rabbit is Rich*, p. 26)

In this radio excerpt, Rabbit relives the memories of his childhood in high school through this invigorating and gentle music. His pleasure is the radio. He likes to listen to it from his car when he drives through the streets of Brewer. The novel becomes a symbolic and emblematic asset that refers to the earthquake, imbalance, dislocation, disarticulation and clash of a text. A real effort is needed to reconstruct the dismantled textual fragments of the novel.

The reader witnesses a textual chaos in the narrative space under the influence of various tones: radio and music. These multiple voices cause a total imbalance that affects Updikian writing on harmony and linearity. Operating on the principles established by the media. Sometimes, we can see that there are more gaps or discrepancies caused by the different media insertions. "The presence of media devices in the novel shapes new ways of reading. The reader becomes more active as a result of the fragmentation of the diegesis." (Atcha 2014, p. 162)

In short, the introduction of the media into John Updike's literary universe encourages the presence of fragmentary texts. The identification of these fragments depends on the type of media present in the body of the text. Each type of textual fragment constitutes a voice that has its realities and participates in the construction of a whole.

### 5.2.1. *Work on the page*

In the Updikian novel, writing is used to cause disorder to any form of established stability. We are witnessing a dispersion of structures that are causing the rejection of determination and unity. In this project of the identity and formal problematic:

> "Writing then emerges as a scriptural rampart against the threat of a negative dissemination of identity components. If always already, through its fragmentary, even divided form, the[narrative] highlights the identity discontinuity of the inscriber 'I', by indicating the date and by respecting the timeline, the[text] is also an object that connects. This means that if the genre allows contradiction, it simultaneously invites respect for a certain linearity of this discontinuity, thus preventing a radicalization of dispersion." (Sergier and Vanderlinden 2012, pp. 7–14)

The medialism in John Updike's novel is perceived as the expression of beauty in writing that is enameled with disparate media entities. This is the exhibition or presentation of different media structures within the textual universe that evokes emotions and sensations, both on a formal and audiovisual level. The inclusion of several types of media in the novel gives it an esthetic allure in disorder. We are witnessing "literary terrorism" as evidenced by this paragraph in *The Brewer Standard* newspaper, in the following excerpt:

> The other news story was about a Washington, D.C., physician who beat a Canadian goose to death with his putter on the 16th

> green of a country club. (The goose honked just as he was about to sink one.) The reason for printing those letters was to demonstrate that truth is stranger than fiction. (*Rabbit is Rich*, p. 271)

In the novel, the font of the textual fragment of the newspaper is 10, while the font of the narrative is 12. This difference in font size already justifies the esthetics of hybridity because Updikian writing steals a certain harmony. The newspaper stands out from the rest of the story.

This separation in form automatically leads to a combination of several structures in the textual landscape that gives John Updike's novel a "rainbow" look. Hybridity, it is this crack in the narrative's fabric that allows the reader to grasp all forms and structures in the narrative journey.

The textual chaos in John Updike's novels is much more noticeable when handwritten. Here, we are rather witnessing an artistic disposition in the work of art. This leads to the fragmentation of the fiction space and the creation of textual clusters of several kinds. One of the most illustrative examples of this kind is Nelson's letter from Colorado delivering a greeting to his parents:

> Hi Mom & Dad & Grandmom:
> These mts. make Mt. Judge
> look sick! No snow tho,
> just plenty of grass (joke).
> Been learning to hang glide.
> Jobe didn't work out, guy was
> a bum. Penna. beckons.
> OK if I bring Melanie
> home too? She could get
> job and be no trouble. Love,
> Nelson
> (*Rabbit is Rich*, p. 37)

By the formal arrangement of the handwritten text in the body of the narrative, the reading offers itself to an esthetic of hybridity that leaves something to be desired through this collage technique. Nelson's letter stands out from the rest of the text narrated by the font. It is not justified as the text of the story. This textual fragment is designed in a style that does not respect the norms of grammar. Instead, we see incomplete, disconnected sentences where English grammar is trampled underfoot. There is also this fragment from *Golf Magazine*, entitled "Annual Roundup" in the following excerpt:

> A COSTLY BIRDIE
> Dr. Sherman Thomas cooked his
> own goose when he killed one of
> the Canadian variety at Congres-
> sional C.C. The court levied a
> $500 fine for the act. (*Rabbit is Rich*, p. 425)

This combination of the handwritten text and the magazine in the novel in an alternating way establishes a dynamic of hybridity in the narrative fabric. The magazine has a title that is in print and bold. The font size used for its writing is 10, which differentiates it from the writing of the novel which is 12. This overall scriptural technique creates a "textual earthquake" in the narrative environment and promotes a narrative of breakage and debris.

The Updikian esthetic favors synesthesia, hybridization and fusion. It is in a dialogue with several entities within the textual universe that the whole philosophy of "beauty" is expressed. Indeed, the anarchic insertion of different kinds of media into the narrative's fabric makes it possible to speak of textual dislocation. Under the effect of the massive proliferation of the media, we are witnessing a certain cadence of literary rhythm, as this other handwritten letter by Janice to her husband Harry Angstrom confirms, in the following excerpt:

> Harry dear-
> I must go off a few days to think. Please don't try to find or
> follow me please. It is very important that we all respect each
> other as people and trust each other now. I was shocked by your
> idea that I keep a lover since I don't think this would be honest
> and it made me wonder if I mean anything to you at all. Tell
> Nelson I've gone to the Poconos with Grandmom. Don't forget
> to give him lunch money for the playground.
>
> Love,
> Jan
>
> (*Rabbit Redux*, p. 91)

This textual fragment offers the reader something pleasant to see by the simple formal arrangement. The reader experiences a particular literary construction process through this letter of separation. There is a variation in style whose scope is subject to the laws of inconsistency and libertine in writing. This part of the story shows a semantic shift. This linguistic transformation creates a sensory effect in the narrative journey.

In addition, the esthetics of hybridity is much perceptible in John Updike's novel through telephone communications. Here, we are witnessing an alternating dialogue between a "transmitter" and a "receiver". The narrative discourse is fragmented into metanarratives and confuses reading. Telephone communication brings joy, emotion and sometimes sadness to John Updike's literary texts. All this gives a certain charm to the writing as can be seen in the passage of the following excerpt:

> The phone rings. Janice plucks it from its carriage on the kitchen wall and says, "Hi, Daddy. How was the Poconos? Good. I knew she would. She just needed to feel appreciated. Of course he's here. Here he is." She holds it out to Rabbit. "For you."
>
> Old man Springer's voice is reedy, coaxing, deferential. "Harry, how's everything?" "Not bad."
>
> "You still game for the ball game? Janice mentioned you asked about the tickets to the Blasts today. They're in my hand, three right behind first base. The manager's been a client of mine for twenty years." (*Rabbit Redux*, p. 80)

The telephone communication takes place first between Janice and her father, then between Janice and Harry Angstrom. She's emotional before she's in love. The reader is thus confronted with triangular communication. There is a distribution of speech between the three characters that gives a linguistic diversity in the novel. This reveals an esthetic aspect in the plurality of voices and the transfer of the affect. The narrative is animated by an audible cadence that evokes a certain emotional feeling in the reader.

John Updike's novels highlight diversity, multiplicity and plurality to highlight the textual hybridity that offers the full artistic dimension to the novel. The different media involved in the story give it a certain color and richness. The narrative is sometimes sung, sometimes viewed, listened to and read. The electronic letters from this motel (see the following excerpt) are sufficient proof of this:

> QUEEN SIZE BEDS
> ALL COLOR TVS
> SHOWER & BATH
> TELEPHONES
> "MAGIC FINGERS" (*Rabbit Redux*, p. 424)

Under the influence of the heterogeneous invasion of media into the textual space, the novel metamorphosis takes on the color of the elements it welcomes within it. The formal hybridity of the book, which also contains screenshots, sounds and images, gives it an identity. All these media elements are used to tell a story in order to give the novel an esthetic value.

In addition, John Updike's novel navigates between different creative processes. The disparate association of media elements in textual structures contributes to creating not only the resonance effect, but also to giving the novel a particular character. We are thus witnessing a kind of "epidemic of the imaginary" that Updike shows by taking care to dispose freely and deliberately of the textual clusters or fragments, as can be seen in this newspaper article reproduced in the following extract:

> *ARSON SUSPECTED IN PENN VILLAS BLAZE*
>
> Out-of-Stater Perishes
> West Brewer police are still collecting testimony from neighbors in connection with the mysterious fire that destroyed the handsome Penn Villas residence of Mr. and Mrs. Harold Angstrom.
> A guest in the home, Mill Jiss
> A guest in the home, Miss Jill Pendleton, 18, of Stonington, Connecticut, perished of smoke inhalation and burns. Rescue attempts by valiant firemen were to no avail.
> Miss Pendleton was pronounced dead on arrival at the Sister of Mercy homeopathic Sisters of Mercy Homeopathic Hospital in Brewer. (*Rabbit Redux*, p. 362)

This article is reproduced verbatim as the one in the novel. Its layout highlights the esthetics of Updikian writing. We have a title in block letters, and the body of the text is rendered in "bold" style. It is in the process of being developed. Hence the many mistakes and repetitions such as: "A guest in the home, Mill Jiss"; "A guest in the home, Miss Jill Pendleton." This article is centered on the page and shifted from the rest of the story.

This formatting of the article in the literary landscape requires a dislocated, fragmented or broken reading. The novel becomes the playground for several creations, associations, organizations, imitations and inspirations. All this phenomenon of foraging exposes "the pleasant" in the textual fabric. This music, these films from various horizons that cross borders, not only immerse the reader in an atmosphere specific to the audiovisual, cinema or written press, but also guarantee a cultural identity for the work.

### 5.2.2. *Media and cultural transfers*

The notion of hybridity, as proposed by Homi Bhabha "in the places of culture makes it possible to understand the plurality of cultural traits involved in constructions and to measure their impact on relational dynamics" (Bhabha 2007, p. 412). Hybridity, for Bhabha, consists of a "third space" where new transcultural identity forms are created. It is also a place of ambivalence rather than a simple and constant opposition. Thus, according to Marie-Hélène Jeannotte:

> "In literature, hybridity also manifests itself, and probably strongly, in textuality: the mixing of forms, registers and languages, as well as the multiplicity of voices or points of view represent some textual marks of a valuation of hybridity that reinforce explicit themes and positioning." (Jeannotte 2010, pp. 297–312)

In John Updike's novels, the perspective of identity is based on a paradigm that permanently associates the digital identity with the identity of nature. This perspective animates the rational project of founding the modern subject and the humanist conception of culture as unitary, homogeneous and stable.

The cosmopolitanism of John Updike's romantic society implies this awareness of belonging to the whole of humanity and not only to his homeland. It is about behaving as a member of the global community and not as a citizen of a state. Thus, under the impetus of the media, we are witnessing a cultural mobility that reconfigures the identity paradigm.

The Updikian texts therefore propose to move from the research of the American novel to a transcultural reading. It is not only the principle of the "nation-state" that is being challenged, but also the mourning of a sedentary, immutable and immobile culture. "Poetics is then understood as an art not only of writing and reading literary texts, but also of recognizing traits, of describing processes that produce the transit effect […]." (Semujanga 1999, p. 28) Zygmunt Baumann describes such writing as "liquid life", i.e. "life caught up in the constant flow of mobility and speed" (Baumann 2005) and which distinguishes it from the fluidity associated with movement.

Indeed, in John Updike's work, the novelistic universe plunges the reader into a cultural diversity under the interweaving of the media. The instability, variation or fluidity of cultures creates an illusion that transcends borders. Cultures interact with each other. Cultural interaction creates a heterogeneous dynamic that constantly updates awareness of identity and difference. The work of imagination consists of opening up to other cultures. This magazine that Rabbit picks up from the table (see the next excerpt) highlights this culture in transit:

> In nervousness he plucks a magazine from the table. It's a Catholic magazine the size of the *Reader's Digest.* He tries to read a story about a lawyer in England who becomes so interested in how legally unfair it was for Henry VIII to confiscate the property of the monasteries, that he becomes a Roman Catholic convert and eventually a monk. The two men whisper together; one maybe is the other's father. The younger one keeps kneading his hands together and nodding to what the older man whispers. (*Rabbit, Run*, p. 195)

This story of the magazine transports or plunges the reader directly into English culture. The reader thus achieves mobility in immobility thanks to the magazine medium. Without him stepping on British soil, English culture was delivered to him. He reads or experiences the same cultural realities as the English. From the reception on the interaction of the text with the audience, the importance of literary history in shaping the narrative process is added. It is important to say here that each culture has its own specificity in itself that distinguishes it from other cultures. The era of globalization thus provides an opportunity for people to open up to cultural diversity.

In addition, this awareness of openness to globalization aims to dominate national values in the name of universal values. Major flows are restructuring the lives of societies around the world: migration and information flows. The reception of ideas and cultures owes much, not only to time, but also to the desire for change. Culture is on display on screens and imposes itself on the whole world in a wealth of diversity since:

> "In Appadurai's view, the cultural flow resulting from globalization is closely linked to the richness of imagination in the different communities of the world. This imagination will remain a prerogative of the post-electronic era. It will also remain a central element in the overflowing modernity, especially since it is without regular consciousness." (Appaduraï and Gustavo 1996)

The author distinguishes two types of imaginations: individual imagination and collective imagination. It is the latter that updates this present post-electronic world. Electronic communication makes the reception of cultural diversity more active and therefore more powerful. The flow of the mass media makes the circulation of information and ideas faster in John Updike's literary landscape. What is global becomes local and vice versa. Characters from different backgrounds communicate with each other through different means of communication.

The media also diffuses several pieces of information in the story. In addition to these pieces of information and images on loop, we also witness a narrative that is sometimes sung and danced in the colors of several melodious sounds that cross borders. In such a semantic field, the reader tends to find an anchor in the flourishing diversity of the world that operates through influence and inter-influence. But beyond such an anchoring, there is a strong desire somewhere to overcome stagnation. This passage from the novel (see the next excerpt) brings to life certain events outside its territory through television:

> "American deaths in Vietnam were reported the lowest in three years, and one twenty-four-hour period saw no American battle deaths at all. The Soviet Union made headlines this week, agreeing with the U.S. to ban atomic weapons from the world's ocean floors, agreeing with Red China to hold talks concerning their sometimes bloody border disputes, and launching Soyuz 6,

> a linked three-stage space spectacular bringing closer the day of permanent space stations. In Washington, Hubert Humphrey endorsed Richard Nixon's handling of the Vietnam war and Lieutenant General Lewis B. […]
>
> In Chicago, riots outside the courtroom and riotous behavior within continued to characterize the trial of the so-called Chicago Eight. In Belfast, Protestants and British troops clashed. In Prague, Czechoslovakia's revisionist government, in one of its sternest moves, banned citizens from foreign travel. […]" (*Rabbit Redux*, pp. 333–334)

This television report takes the reader from Vietnam to Prague (Czechoslovakia), Washington, the Soviet Union, China, and back to Washington, Chicago and Belfast. The story thus provides an opportunity to travel around the world in a state of immobility. In this mental journey, we are witnessing, for example, the Vietnam War, agreements for the prohibition of the use of atomic weapons between the United States and the Soviet Union, not to mention riots outside a courtroom in Chicago. This mobility of culture creates a panoramic vision of the world for the reader and at the same time imposes the mixing of cultures. In this way, the question of cultural identity in a global context is raised. The unifying principle of cultural diversity thus lies in the power to engage in dialogue with a view to finding a platform that ensures true stability. Globalization erases differences and imposes cultural openness.

## 5.3. Conclusion

The John Updike novels analyzed in our study are distinguished by a strong or exaggerated invasion of media that exposes the story to a chorus of media voices and a media description of fragmentation. This narrative dynamic combines on-screen literature with a telephone story that offers the reader a "give-and-take" in the textual landscape.

The Updikian novel is thus at the crossroads of a medialism perceived as "an exacerbated or completed form of intermediality [i.e.], writing that is one with the media, [where] strictly literary and media materials [participate] together in the literary nature of the text" (Dého 2014, pp. 186–187). This literary writing of the media, as demonstrated by the structures of

*Rabbit, Run; Rabbit Redux* and *Rabbit is Rich*, creates an illusion that transcends borders and promotes not only cultural mobility, but also dialogue between cultures.

Cultural interaction thus creates a heterogeneous dynamic that constantly updates awareness of cultural identity and diversity. With the anarchic involvement of these media, the texts sometimes allow us to talk about the esthetics of a hybridity that offers "charm" or "pleasure" with sensory effects in a climate of "living together".

## 5.4. References


Appaduraï, A. (1996). *Modernity at large: Cultural Dimensions of Globalization*. University of Minnesota Press, Minnesota.

Atcha, A.P. (2009). Pratique intermédiale et création romanesque chez Williams Sassine. *En-Quête*, 21, 49–65.

Atcha, A.P. (2011). *L'écriture postmoderne dans le roman ivoirien. Le postmodernisme dans le roman africain: formes, enjeux et perspectives*. L'Harmattan, Paris.

Atcha, A.P. *et al.* (2014). *Médias et littérature: Formes, pratiques et Postures*. L'Harmattan, Paris.

Bakhtine, M. (1970). *La Poétique* de *Dostoïevski*. Le Seuil, Paris.

Baumann, Z. (2013). *La vie liquide*. Fayard, Paris.

Bhabha, H. (2007). *Les lieux de la culture. Une théorie postcoloniale*. Payot et Rivages, Paris.

Dého, T.R. *et al.* (2014). *Médias et Littérature: Formes, pratiques et postures*. L'Harmattan, Paris.

Deslauriers, J.P. (1991). *Recherche qualitative. Guide pratique*. McGraw-Hill Education, New York.

Garrigues, P. (1995). *Poétiques du fragment*. Klincksieck, Paris.

Guiyoba, F. (2015). *Littérature médiagénique: Écriture, musique et arts visuels*. L'Harmattan, Paris.

Jeannotte, M.-H. (2010). L'identité composée: hybridité, métissage et manichéisme dans la saga des Béothuks, de Bernard Assiniwi, et Ourse bleue, de Virginia Pésémapéo Bordeleau. *International Journal of Canadian Studies*, 41, 297–312.

Mariniello, S. (2003). Commencements. *Intermédialités*, 1.

Méchoulan, É. (2003). Intermédialités: le temps des illusions perdues. *Intermédialités*, 1, 9–27.

Müller, J.E. (2010). Intermediality and Media Historiography in the Digital Era. Acta Univ. *Sapientiae, Film and Media Studies*, 2, 15–38.

Prince, G. (1978). Le discours attributif et le récit. *Poétique*, 35, 305–313.

Rajewsky, I. (2005). Intermediality, intertextuality and remediation: A literary perspective on intermediality. *Intermédialités*, 6, 43–64.

Semujanga, J. (1999). *Dynamique des genres dans le roman africain: éléments de poétique transculturelle*. L'Harmattan, Paris.

Sergier, M. and Vanderlinden, S. (2012). Le journal d'écrivain. Les libertés génériques d'une pratique d'écriture. *Interférences littéraires/Literaire interferenties*, 9, 7–14.

Updike, J. (1960). *Rabbit, Run*. Alfred A. Knopf, New York.

Updike, J. (1971). *Rabbit Redux*. Alfred A. Knopf, New York.

Updike, J. (1981). *Rabbit Is Rich*. Fawcett Crest Books, New York.

Van Spanckeren, K. (1994). *Esquisse de la littérature américaine*. Africa Regional Services, Paris.

PART 2

# Issues Linked to the Digital Economy in Africa

# 6

# "*Je suis au bara*" or the Deviant Use of the Internet among Abidjan's Scammers

Scammers are specialists of an endemic form of cybercrime focused on Internet scams. In Côte d'Ivoire, scamming is a serious social phenomenon in which the State[1], the media and several scientific fields such as criminology, sociology, law, economics and communication are interested. While most scientific studies focus on explanatory factors, manifestations, solutions and socio-economic impact (Bogui 2009, 2010; Gueu 2013; N'Guessan 2014), the legal framework adapted to the complex nature of cybercrime (Mitongo-Kalondji 2010; Tanon-Bian 2015), very few of them specifically address scamming. This chapter therefore contrasts with these holistic studies and argues that understanding this phenomenon requires an investigation involving scammers. As a result, it gives a direct voice to the scammers who are rational individuals with a share of autonomy, freedom that combines means, they anticipate and evaluate alternatives according to their goals, intentions and expectations. A scammer therefore is an active individual who uses ICTs as an extension of existing social practices such as crime (Jouët 2000). Studying these do-it-yourself uses (De Certeau 1990) from an explanatory and individualistic perspective makes it possible to grasp

Chapter written by Ettien Franck Stéphane ADOU and Mahesse Stéphanie KOLÉ.

1 The fight against cybercrime in Côte d'Ivoire is based on the one hand on the Ministries of ICT, Interior and Justice, and on the other hand on ICT regulatory institutions such as the *Direction de l'informatique et des traces technologiques* (DITT) and the *Laboratoire de criminalistique numérique* (LCN), the *Plateforme de lutte contre la cybercriminalité* (PLCC) and the *Autorité de régulation des télécommunications en Côte d'Ivoire* (ARTCI), as well as Act No. 2013-451 of June 19, 2013, on the fight against cybercrime, journal officiel de la République de Côte d'Ivoire.

the individual logic and social constraints involved and understand scamming. Therefore, understanding this phenomenon requires an understanding of what they themselves attribute to their practice.

In this context, the work on the 419 Nigerians, the Cameroonian *Feymen*, the Beninese *Yahooboys*, the Ivorians (Koenig 2014; Tasso-Boni 2014; Adou 2016) is of interest to us. Indeed, they operate a paradigmatic reversal by studying male "deviant uses" (Adou 2016; Bahi 2016), "intelligent and innovative" uses (Tasso-Boni 2014) of ICT as well as the social practices of a category of ICT users stigmatized and labeled as deviant (Adou 2016). Scammers appear as people whose ways of doing, thinking and living are different from the rest of the population. They are characterized by a diverted and transgressive use of ICT. By using ICT as a scam tool, they are alienating themselves from the prescribed uses of ICT, social values and legal norms that govern society. This is referred to as "deviant use".

Scammers constitute an environment of inter-knowledge (Beaud and Weber 2010) insofar as they know each other by name, by reputation, frequent the same places and share the same logic, goals, uses and practices. They constitute a homogeneous set of stigmatized and labeled actors within which know-how, skills and "segments" of expertise circulate ((Baszenger 1990); Strauss cited by (Delas and Milly 2005)). The tortuous nature of the practice has fostered the emergence of an "informal code of ethics" and greater peer review. Shared uses, beliefs, values and representations structure this practice, which is for some people a means of subsistence. We therefore maintain that scamming is a profession in its own right for scammers.

The concept of profession from a functionalist approach can be summarized using three criteria: "the specialization of knowledge that leads to a precise and autonomous determination of the rules of activity; higher-level intellectual training that presupposes the existence of duly recognized training schools; an ideal of service that requires the establishment of a code of ethics and peer review" (Boudon *et al.* 2012). This definition limits professions to those established with standards and codes that structure them (doctors, lawyers, architects). It does not consider liberal professions, small trades, let alone deviant or criminal practices such as scamming, which share fundamental characteristics with established professions. The interactionist approach is better adapted to the criminal and secret nature of scamming. It views professions as a generic form of occupation and identity in the work environment. In its broadest sense, the term profession is synonymous with a trade and refers to an activity from which the actor can derive a livelihood. A profession also characterizes all individuals who perform the same occupations in the production process of a good, service or practice. Thus, understanding scamming as a profession from an interactionist approach gives rise to reflections on how the scammer presents and represents himself through their practice. Therefore, a series of more specific questions are needed: who are these scammers? How does one become a scammer? How is know-how transmitted or acquired? What are the actual uses and practices developed? What are the routes and careers specific to the practice? What are their rewards? How can this practice be considered a profession?

At the heart of these questions is a set of anthropogenic concerns based on access logic, modes of transmission, status, effective uses, practices and the identity of scammers.

## 6.1. Materials and methods

This chapter adopts the anthropology of communication approach (Winkin 2004; Bahi 2016). It uses data from an ethnographic survey we conducted for eight months at Cybercafé Pro, an Internet café in Abobo-Abidjan (Adou 2016). Internet cafés are the preferred place where scammers gather and get to work and engage in "repeated personal interactions" (Beaud and Weber 2010, p. 32). The cybercafé is both a place of connection and rallying and a place of work. This socio-anthropological survey was based on participant observation *in situ*, comprehensive one-on-one interviews and group discussions with members of the community from whom the Internet café was visited. Based on the themes, occurrences and specificities that emerged through a longitudinal content analysis, we have managed to categorize access logic, transmission modes, status, effective uses and scamming practices. It should also be noted that this article mobilizes the reflections of a study on the profession of communication managers in Côte d'Ivoire (Kolé 2016) as well as the concepts of the interactionist sociology of Howard Becker (1985) on deviance (labeling, deviant career, deviant subculture), Everett Hughes and Isabelle Baszanger (1990). Claude Dubar's (1999) reflections on the process of emergence and legitimization of professions have also been taken into account. Typical italicized expressions and conceptual reminders are used as the work progresses.

## 6.2. From the robot launcher to scamming master: routes and upward mobility of the scammer

The *robot launcher* is one of the first levels of expertise in the practice of scamming. It is a subordinate stage which, after years of training, prefigures the stage of "*scamming master*", the ultimate stage of success. Scamming is a sequential phenomenon (Becker 1985) which implies access logic, modes of transmission of know-how and changes in status. It should be noted that Cybercafé Pro browsers use the term "bara" more often to describe their activity. In Dioula, it refers to "profession, work" and the verb "to work". The words *je bara* (I work), *je vais au bara* (I go to work) and *je suis un barasseur* (I am a worker) therefore refer to the practice of scamming.

### 6.2.1. *Access logic and scamming transmission mode*

Anyone who wants to learn scamming approaches an experienced scammer for training. All Cybercafé Pro browsers expressed a desire to learn by going voluntarily and deliberately to those who could introduce them to the profession. Thus, by cross-checking access data, it can be seen that those who learned *bara* around 2006 were trained directly by Nigerians, or indirectly by the students of these Nigerians, in South Abidjan (Marcory, Koumassi, Port-Bouët). However, from 2008 onwards, learning is also taking place in the working-class districts of Yopougon and Abobo. The masters are no longer Nigerians but rather experienced friends and also Internet café managers. In any case, this know-how is transmitted in the cybercafé through practical training and especially in contact with the most experienced people. In this context, there are relationships of hierarchy and subordination between "a master trainer", an "old father", a *kôrôh*[2] and an aspiring scammer, an apprentice, a "good little one". By rubbing shoulders with the masters and participating in their scams, apprentices learn bara.

The apprentice is a beginner who wants to learn, who is introduced to a trade. As an apprentice, they have to accompany their "old father" to their *chahuts* (parties) and buy them cigarettes and food. These tasks are marks of respect, submission and determination that demonstrate their willingness to learn. These are the following symbolic acts of an initiatory nature that bind

2 "Old father", or *kôrôh*, refers to an older person to whom respect is due. They can extend to the respect that is due to a person, because of their talent or experience in a field.

and mark the apprentice's allegiance to the trainer's know-how. Thus, the apprentice gradually acquires their know-how by working with their master.

The trainer, the master, or the "old father", receives respect from the apprentice, on the one hand, because of their birthright and, on the other hand, because of the know-how they pass onto them. The learning relationship is the main source of acquisition and transmission of techniques. This relationship is also present in traditional societies where the transmission of knowledge was mainly through initiation. This relationship is important for apprentices who are at the same time introduced into their new group. The "old father" is a moral guarantor without which it would have been difficult for the "good little one" to join the group. From that moment on, they benefit from the experiences of their predecessors. They provide them with the know-how that will allow them to achieve their objectives with minimum issues. They give them the solutions adapted to the problems they will probably encounter in terms of choosing victims, operating methods and withdrawing money.

### 6.2.2. *Phases, success models and rewards of the scamming career*

The phases refer to changes in state, status in the evolution of the scammer's career. These are the different constitutive, successive and distinct stages of a profession, of an activity. They represent the different steps in the evolution towards a model of success in the profession. The transition from one phase to the other involves the acquisition of knowledge, attitudes and skills. We use Howard Becker's (1985) phase division in his study of marijuana smokers[3]: beginners, occasional and regular.

The novice scammer must first learn the codes and procedures of their profession. They are very often the one who stands up in the Internet café, observe and question the activities of those who are in the *process of bara*. It is at this level that we find the *robot launcher* who *hunts* and searches for targets for the most experienced scammers. This is the first expertise acquired

3 Just like marijuana smokers who were the subject of the study, scammers are exposed to the same processes of stigmatization, disqualification, and labeling. Becker's study appears as a relevant framework for reflection to analyze individuals and groups considered by society as outsiders, deviants and go-betweens.

and can be considered as the first step. It is housed in the operational level and has no decision-making power. They are performers who comply with the profiling requirements determined by the sponsors. At this level, the borrowing of specific vocabulary, clothing style, norms and values can be analyzed in terms of identification and progressive integration into the reference group. Beginners are mainly young students and out-of-school people who are older. They all know how to read and write. Out-of-school beginners spend more time in the Internet café than out-of-school beginners. While most beginner students only do it in their free time (at noon, in the evening after school, on weekends, during holidays), the unschooled beginner will spend more time in the Internet café. This time factor determines for many the transition to the second identified phase.

The occasional scammer is one who has already completed at least one operation or is trying to do so. They have already learned the basics of operating methods and techniques (blackmail, stalking, withdrawal of money). They are able to trace the different steps of an operating procedure up to the withdrawal of the money. However, they remain dependent on their relationship with the person who has introduced them to the profession since they continue to be trained. Some occasional scammers have another activity to which they devote time (football training club, vocational training, general education) or which brings them money (sale of clothing). In this second phase, the scammer acquires a broader view of the operating methods. They can therefore vary their techniques, depending on the victim and the amounts they want to obtain. Their work is located between operational and strategic. They ensure the design and implementation of the strategy but always receive guidance and advice from their master or a more experienced friend. From this stage, scammers carry out several activities together: playing football, going to the maquis (bar). They are part of a process of isolation and self-segregation (Becker 1985) and move away from the community of non-scammers.

A regular or "patented" scammer is one who devotes themselves exclusively, systematically and routinely to scamming, which becomes the main activity around which they organize their way of life. Thus, re-socialized by the group, this individual develops typical motives and interests and internalizes internal and external behavioral norms (clothing style, pursuit of money and contempt for specialized institutions fighting cybercrime).

The concept of career emphasizes the reference and the professional ideal that members of a profession want to achieve, through the notion of success models (Becker 1985). It should be noted that this is the representation of success by the scammer and may therefore be different from the success models of other professions. Indeed, Cybercafé Pro's scammers identify with singers, athletes or other scammers. They have in common that they have money, material goods and are famous. Scammers, on the other hand, are evaluated by the amount of scams called the *West*[4]. Success is also sequential and proportional to scammers' ability to regularly succeed in scams, the amounts of scams accomplished and the use they make of the money earned. Success is measured by attendance at nightclubs and acquisition of "luxury items" such as cars, motorcycles, luxury watches and the latest-generation mobile phones. These "luxury items #" and night trips are the manifestations and the indicators of this success that also consider the *bara*'s skill, reputation, prestige and esteem of the reference group.

The first level of success is the one where the scammer occasionally saves small amounts of money. They can fund their customer account at the Internet café, buy new clothes and host some *chahuts* (parties). At this stage, they know the famous places, but only frequent them depending on the success of their scams. However, they are proud and happy to fully carry out scams up to completion. The second level of success is the one where they go out every weekend to air-conditioned clubs and bars. At the third level of success, they become a glory of the profession, *scam masters*. Indeed, the scam master has a lot of money and shows it. Their extravagance and their ostentatious wealth are talked about. They spend without counting because they are not afraid, "their hearts do not break". They are known to almost everyone in the industry. They organize and sponsor parties where they show that "they have the money". Their nickname appears in spots[5] by famous artists. They are completely autonomous since they rent houses, have cars and work only for themselves. They are at the top of the pyramid of success among scammers.

4 With reference to the Western Union international money transfer system, the term *West* refers to the money obtained as a result of a successful scam. It is most of the time through this transfer system that victims send the money.

5 The spot is a variant of the "Coupé-Décalé", which only focuses on praising those who have paid for it. In the spot, the person's nickname, qualities and values are exalted.

Scammers attach emotions and strong feelings to their practice leading to an impression of satisfaction and rewards. There is a sense of power and strength linked to the possession of money, the desire to belong to a group, to a "family" and to a privileged class of "rich young people" who constitute rewarding prospects. Scamming appears as a liberal profession that allows a person to have a clothing store, travel abroad and offer themselves everything they want. All these elements represent a source of pleasure and satisfaction that the scammer gradually gets to know. Learning about this joy is done through contact with the most experienced scammer. They make people aware that this practice is likely to bring a lot of pleasure to those who indulge in it and that they can even live off it. Going out to the "maquis" and bars with peers allows them to enjoy what they do. These interactions reassure the novice scammer of the rewarding perspectives, and this awareness determines the progression in practice depending on whether one has appreciated and whether or not one is seeking the advantages and disadvantages of this new life.

## 6.3. The scamming practice

Scammers operate and interact in the Internet café, which is not only a meeting place but also a workplace. The aim is to highlight the actual practices taking place during a typical *bara* day by focusing on the techniques used.

### 6.3.1. *Being in the bara: a typical day at the Internet café*

Cybercafé Pro offers Internet connection as its main service. It is a cramped and crowded place, mainly frequented by bulldozers and a few "ordinary customers". Scammers are recognizable by their young age, personal appearance, nickname and behavior. Their age varies between 12 and 25 years. They have a casual look, clothes with bright colors and almost always the same blue sandals with white stripes. They only use nicknames and always look happy. They are noisy and this is one of the reasons for the conflicting nature of the interactions between the scammers and the manager of the Internet café.

The scammers take over the Internet café by listening to music, chatting on the terrace and in the Internet café: they make themselves comfortable for *bara*. The manager criticizes them for their lack of manners that she understands as rudeness. Therefore, insults are hurled and verbal fights occur regularly. Cybercafé Pro is very popular, but has a variable attendance that is explained by the fact that scammers organize their attendance according to their real activity (student), the virtual activity of their avatar and the activity of their target. However, there is a peak in the morning and afternoon from 4 p.m. onwards. Within the bara, we observe that some scammers throughout the day always and almost use the same computer they "privatize". In addition, they are constantly in communication or gather around an individual who works on the computer. This refers to either tandem work or training. We speak of sedentarization when it is the same individuals who use Cybercafé Pro.

### 6.3.2. *Processes mobilized within bara: between psychology, technology and mysticism*

The skill inventories can be assimilated to the operating procedures developed during scamming. These are the different techniques that Cybercafé Pro scammers use or are likely to use to scam. They are based on expertise, psychotechnical and mystical processes.

Psychotechnical processes are strategies that require mastery of the computer tool through the sending of photos, videos or avatar creation, knowledge of the target's psychology and also acting skills to imitate the people they claim to be. It is not enough to know the techniques of scams, but it is also necessary to adapt them to the time and target. Thus, based on the processes developed, two categories (not exclusive) of scammers were identified: "perverts" and "lovers".

Perverts mainly use video blackmail. This technique allows the scammer, in possession of a video showing the victim in compromising or humiliating positions, to extort money from them. For this technique, the victim is accused of pedophilia and administrative documents (arrest warrants, passports) are prepared in order to make the threats credible. Lovers, on the other hand, develop a scam of false feelings. By inventing dramatic situations, the scammer encourages the target to react out of compassion, pity or love. The aim is to create a real feeling in the victim based on false stories. These seemingly simple techniques require preparatory work since, in addition to pseudonyms, the victim's identity (surname, first names, profession) must be provided. To do this, it is necessary to maintain ongoing relationships to gain the trust of the victim who will get used to it and provide all of their personal information. Scam processes are dynamic and adapt to the victim's spatial and temporal context, mentality and psychology. The secret of scamming is to identify the psychology, the *mind* of the victim, in order to predict the possible alliances to be woven and the particular expertise to be mobilized.

A profession is based on skills and expertise. Expertise consists of the production of action-oriented specialized knowledge. Its legitimacy is based on the success of their actions and competence. The expert is recognized as such among other professionals in their field. They derive their competence from the mastery of specific knowledge and their own experience (Trépos 1996, pp. 4–10). And in scamming, expertise is strategically exploited. Indeed, Cybercafé Pro's scammers build relationships with other scammers who specialize in hunting (robot launchers), avatar creation, official and administrative document production. Scammers need to have relationships at the level of financial institutions for the withdrawal of money. This expertise is necessary to carry out the scams. The avatar design scammer creates a profile corresponding to the aspirations of the targets previously identified on social networks by a robot launcher. In this case, everything suggests that the scam is indeed (place, work, social network account) where it claims to be and is who it claims to be. They have videos and photos to reassure the customer. Some scammers are specialists in the preparation of official documents intended either to make threats credible (arrest warrants, inheritance blackmail) or to authenticate sales (notarial deeds in the sale of land, houses or businesses). Scamming is based on psychotechnical processes, specializations, as well as on mystical practices: Zamou.

*Zamou*[6] is a term used by scammers to describe the use of mystical and occult practices to facilitate scams. They explain that the *Zamou* is the fantastical or mystical process that causes the victim to lose their common sense, their discernment, their logic. A person who is "zamored" is unable to analyze the situation objectively and reasonably because they are under the influence of occult powers. *Zamou* involves making sacrifices in proportion to its intensity. These sacrifices can range from chickens to a part of the person's body (a finger, toe or sexuality). However, in extreme cases, human blood or specific parts of a human body are required. The rituals and

6 Unfortunately, we are not in a position to clarify the etymology and origin of the term *Zamou*.

prohibitions of this practice are diverse: spending the money earned daily, having sex with a child, a madwoman or a dead person.

Many scammers systematically use this process, which guarantees greater gains and *West*. They all recognize the existence and effectiveness of this "mystical supplement". They designate renowned localities in Côte d'Ivoire where it is possible to do this mystical work. There is the city of Dabou[7], more precisely the villages of Gbougbôh, where the Beninese are the mystics and Akradjo, where there is an "old woman who does a good job". The city of Bondoukou[8] is also regularly mentioned. Others prefer to go to Benin, Nigeria or Senegal, where there are great mystics. It is a very common practice, and scammers pass the contacts of these mystics between friends.

The mystical dimension is inseparable from the scamming exercise[9]. This is a "normal" practice that many scammers admit to using to a lesser extent. However, the one who uses it exclusively "without even knowing how to type" and goes as far as human sacrifice is a counter-model of success, because most of them argue that there is no merit in sacrificing someone for *money*. The esteem that a scammer holds must be linked to their ingenuity and to their *mind*. Some argue that those who exclusively use *Zamou* should not even be considered as scammers. They are usurpers who give bad press to their practice. From then on, anyone who performs *Zamou* without a bara or who sacrifices a man is considered by others as a deviant. There is then a profound deviance because this scammer is subject to double labeling: first, that of the global culture, and second, that of the scamming subculture to which they are supposed to belong.

## 6.4. The dynamic identity of scammers

Identity is dynamic. It takes into account the image that scammers have of themselves, their practice, their entourage (friends, relatives, moral entrepreneurs), and it is built in tension with the identity that society attributes to them.

7 Dabou in the South. Akradjo and Gbougbôh are villages in Dabou.

8 Bondoukou is in the northeast.

9 Manipulating a computer.

### 6.4.1. *Reflective identity (self-image and practice)*

Cybercafé Pro's scammers are openly presented as scammers, *barasseurs* (workers), *keyboard typists*. They recognize it without fear and with great pride. The elders proudly proclaim that they are dedicated exclusively to this activity. They present themselves as licensed or confirmed because they manage to make a living from it.

The experienced scammer then appears as a strategist in the "Crozierian" sense (Crozier and Friedberg 1992). They adopt strategies that aim to optimize or maximize their personal gains. The strategist makes strategic decisions and closely directs the execution of their decisions. They constantly adjust their strategy to the new data they are confronted with. They design a strategy and translate it into a plan with friends, their "good little one" or alone. They adapt their strategies according to the context, situations, people involved and the target's environment. Thus, they are perceived as a strategist who feels different from other members of society who do not know what it is about when it comes to scamming. They belong to an "extraordinary" group of people, an elite. They feel smarter because they know how to "enjoy life" better and they succeed in deceiving their victims who do not understand the techniques used or their life choices. They are ingenious and make their *minds* work. This feeling of superiority makes scamming unique and reflects the scammer's rewarding image of himself and their activity.

According to scammers, scamming is not a game, a hobby or a whim, but rather a job: a job in its own right that offers rewarding opportunities to those who engage in it. The scammer cannot be disturbed when they are working (in *bara*). They consider the Internet café as their workplace, their office where they have daily schedules and activities to perform. Scamming is therefore a profession which, moreover, is not as easy as one would like to make believe. The scammer is not an idle person, contrary to the idea that the *doxa* is a vehicle. They decide to devote himself to scamming, which, like other professions, involves sacrifice, respect and merit. Especially since they manage to live off it despite the obstacles. They consider their activity as "the hardest job in the world" which requires discipline to learn techniques, courage to continue despite the criminalization of the phenomenon, creativity in the face of obstacles and above all ingenuity to convince the victim.

Scammers are part of a process of idealizing what they do and who they are. They reaffirm, through the idealized image they give of their activity, that they work as much as members of other professions. They enjoy more rewards. In addition to reflective identity, it is important to focus on the views they have of the people with whom they interact.

### 6.4.2. *The scammers and their entourage*

Scammers identify themselves with the same group that they value and idealize. However, at the same time, they devalue and discredit external categories. They proceed to a social re-categorization by relying on the views they have of their parents, friends, victims and those who oppose their activity, "moral entrepreneurs" (Becker 1985, p. 171). They feel different from their victims, whom they often despise. They assign qualifiers to them according to the context or type of process developed. They believe that their victims are perverts in the case of video blackmail, or "bulls", when the victim is difficult to scam, *mougou* when the victim has been scammed several times. *The* victim may also be a client with whom they have a business relationship. However, in any case, the victim "is in a 'genjutsu'"[10], i.e. an illusion built by the scammer.

Moral entrepreneurs are those who pass laws, substitutes for cybercrime control bodies and conventional law enforcement forces (police). The agents of the *Plateforme de lutte contre la cybercriminalité* (PLCC) "spoil their business". Police officers are called corrupt people who change sides according to the interests at stake. For them, the state authorities (the Interior and ICT Ministers) are "cowards", since it is under the injunction of European countries that they take coercive measures. They think that politicians are the "real scammers" because they steal from the State.

As for the non-scammers, the scammers call them "innocent". The innocent is at the opposite end of the spectrum from the scammer who develops strategies and takes advantage of all the opportunities available to them to get money. The innocent is the one who behaves in a conventional manner and meets all societal standards. The innocent does not know the joys, pleasures and possibilities of scamming. The innocent is the opposite of their

10 Expression from a Japanese cartoon (manga) called *Naruto*.

conception of the "good scammers", since they prefer "to live like a gazelle and not like a lion"[11].

Analysis of the scammers' reactions to the social reaction reveals that they are isolating themselves. Indeed, isolating oneself is a reaction of scammers to social condemnations and the labeling process to which they are subjected. They want, by this technique, to protect themselves from the judgments, condemnations and quibbles of their entourage. Isolation, or the process of self-segregation, is the acquisition of dress, speech and behavior typical of the scamming group, but different from that of individuals in the global society. They are therefore developing techniques for neutralization or self-justification. They are elaborate systems of justification that make it possible to give reasons for one's actions, to feel more comfortable in one's activity because they partially or completely remove themselves from the accusations made by society. This justification system allows scammers to find reasons to continue scamming and to remove any remaining reluctance they may have to carry out their actions.

## 6.5. *Discussion: scamming, a persistent deviant profession*

The term "profession" has at least four meanings: a declaration (professional identity), a job (professional classification), a trade (professional specialization) and a function (professional position) (Dubar 2000). The combination of these elements makes it possible to demonstrate that scamming is indeed a profession. Indeed, one of the indicators of the existence of a profession is the presence of a professional identity. Identity is what defines us and distinguishes us from others through our uniqueness and characteristic attributes. Identity is characterized by a definition of self by oneself and self by others. In an interactionist approach, the professional identity of scammers would result from the interactions developed in the context of scamming. For Claude Dubar, professional identities "are socially recognized ways for individuals to identify with each other in the field of work and employment" (Dubar 2000, p. 95). The construction of this professional identity specific to scamming is based on the articulation between the identity attributed by others (objective identity) and the identity

11 Scammers here divert a proverb that is widely used in their environment and that justifies the fact that they do not care about tomorrow; "it is better to live one day like a lion than every day like a gazelle" justifies that they do not think about tomorrow.

conferred by oneself (subjective). Professional identity is a self-conception at work, a dynamic, continuous and interactive process that is built over the course of a professional life. It is subject to the evolutions and changes imposed by the environment and by the individual himself. Each profession has an objective definition common to all members of society. The individual therefore identifies himself with the activity, with the work they do. Thus, it is the combination of their knowledge, personal style, strategies and skills in their activity that constitutes their professional identity. In short, professional identity is the result of specialized training, experience, choices, habits, encounters and sharing between scammers. This identity is based on professional logic, a career, unified practices and a culture that is typical of them.

Scamming is seen by those who scam as a job, a means of subsistence. In the context of scamming, we observe specificities resulting from interactions in work situations that we assimilate to professional logic. Indeed, when an individual belongs to a professional group and, in so doing builds their professional identity, they share the socio-professional representations of the profession and establishes discriminatory relations with the other groups in their environment (Moliner 1996). The practice of the scamming profession is underpinned by the secret and the "Coupé-Décalé" ethos (Adou 2016, Kolé 2020).

Scammers are considered outsiders, different, deviant. As a result, they are labeled and stigmatized. They appear as a group of people of diverse origins who trace the contours of a criminal practice and work environment. In this sense, it clearly establishes a uniformity in the conditions of access and exercise, and therefore a specificity at the level of representations. Analysis of actual practices reveals that scammers standardize their social or professional rhetoric (Walter 1995, p. 12). Thus, in the different situations in which they operate, scammers have the same practices through which they identify themselves. They are aware that it takes special skills to practice their profession. From the point of view of training, it should be noted that, regardless of how they access the function, all scammers follow the same constituent sequences, the same career. Indeed, like Everett Hughes and Howard Becker, we convene the career concept to analyze the trajectories, statuses, sequences and routes taken by a scammer . The notion of career is understood as "the continuation of the transitions from one position to another accomplished by a worker in a professional system" (Becker 1985, p. 47). It allows us to study typical phases, the psychological changes that accompany each of the stages, the models of success and the reasons for satisfaction attached to the practice of scamming. The existence of a typical career in practice reinforces the idea that scamming is a profession.

However, beyond the trajectories followed by the individual within the profession, an organized set of meanings, conventional behaviors and even specific values is developing. This suggests that the members of this group belong more than to a group; they have a common culture. This culture is distinguished by the set of norms and values that underlie and drive actions. Indeed, culture is "all you need to know to be part of your family, your society, an institution, etc. It is a set of rules, more or less implicit or explicit, that allows human relationships" (Winkin 2004). Howard Becker, quoting Everett Hughes about the concept of culture, states that "a culture is formed whenever a group of people lead a partly common existence, with a minimum of isolation from others, a common position in society and perhaps one or two enemies in common" (Becker 1985, p. 104). It appears that scammers, in the course of their activities, face the same concerns and penalties; they suffer from the same stigmatization. Given the difficulties they face, they tend to regroup and isolate themselves. The phenomenon of self-segregation of scammers therefore reflects the existence of a culture within scammers. As soon as there are ways of thinking and acting specific to scammers, we can affirm the existence of a group culture.

## 6.6. Conclusion

Scamming is a profession that is based on shared professional logic, a typical career, a collective identity and a culture specific to scammers. These elements contribute to the construction of a typical professional identity that emerges from the practice of this activity. However, scamming is a criminal practice that has been condemned since its introduction in 2002. It therefore appears as a "persistent deviant profession" that involves scammers, experts, professionals in the deviant use of ICTs in Côte d'Ivoire. Scamming also appears as a "secret" formation society where knowledge is acquired through an initiatory process. This formation perpetuates the existence of the phenomenon and gives the "insiders" a power, a know-how that makes them different. It builds both the physical, the personality, the identity, the social life, the time and the world of the one who practices it. The scammer thus becomes an element of this whole, a member of this society, of this subculture. The introduction to scamming is an interesting perspective for further study.

## 6.7. References

Adou, E. (2016). Le "broutage" comme usage déviant d'Internet en Côte d'Ivoire: Étude ethnographique de brouteurs dans un cybercafé d'Abidjan. Master's thesis, University Félix Houphouët-Boigny, Abidjan.

Atchoua, N'G.J. and Bogui, J.-J. (2016). La régulation des TIC en Côte d'Ivoire: entre identification et craintes de profilage des populations. *Tic & société*, 10(1).

Bahi, A. (2004). Usages d'Internet et logique d'adaptation sociale des jeunes. Une étude dans des "cybercafés" abidjanais. *Bulletin du CODESRIA*, 1/2.

Bahi, A. (2016). *Anthropologiques de la Communication. Pour une approche anthropologique de la communication et des rapports sociaux.* Langaa RPCIG, Bamenda.

Baszanger, I. (1990). Émergence d'un groupe professionnel et travail de légitimation: le cas des médecins de la douleur. *Revue française de Sociologie*, 31, 257–282.

Beaud, S. and Weber, F. (2010). *Guide de l'enquête de terrain.* La Découverte, Paris.

Becker, H. (1985). *Outsiders. Étude de sociologie de la déviance.* Métailié, Paris.

Bogui, J.-J. (2009). Usages et appropriation des TIC par les jeunes Ivoiriens: de l'espoir au désenchantement. *TIC et Développement, le Net et ses démons*, 18 November.

Bogui, J.-J. (2010). La cybercriminalité, menace pour le développement. Les escroqueries sur Internet en Côte d'Ivoire. *Afrique contemporaine*, 2(234), 155–170.

Crozier, M., Friedberg, E. (1992). *L'acteur et le système.* Le Seuil, Paris.

de Certeau, M. (1990). *L'invention du quotidien 1. Arts de faire.* Gallimard, Paris.

Delas, J.-P. and Milly, B. (2005). *Histoires des pensées sociologiques.* Armand Colin, Paris.

Dubar, C. (1999). La sociologie des professions face à la médecine. *Sciences sociales et santé*, 17(1), 35–39.

Dubar, C. (2000). *La socialisation : construction des identités sociales et professionnelles*. Armand Colin, Paris.

Gueu, D. (2013). La cybercriminalité à Abidjan, un phénomène de mode ou une nouvelle guerre contre les finances en Côte d'Ivoire. *European Scientific Journal*, 9, January.

Hughes, E. (1996). *Le regard sociologique*. Éditions de l'École des Hautes Études en sciences sociales, Paris.

Jouët, J. (2000). Retour critique sur la sociologie des usages. *Réseaux*, 18(100), 487–521.

Koenig, B. (2014). Les économies occultes du "broutage" des jeunes Abidjanais: une dialectique culturelle du changement générationnel. *Autrepart*, 3(71), 195–215.

Kolé, M. (2016). Les responsables de communication en Côte d'Ivoire: vers la constitution d'un champ social. Master's thesis, Université Félix Houphouët-Boigny, Abidjan.

Kolé, M. (2020). Musique urbaine ivoirienne et construction d'identité collective: le Coupé-Décalé comme mode d'expression d'une identité dégagée et désengagée. PhD thesis, University Félix Houphouët-Boigny, Abidjan.

Moliner, P. (1996). *Images et représentations sociales*. Presses universitaires de Grenoble, Grenoble.

N'Guessan, A. (2014). La pratique de la cybercriminalité en milieux scolaire et universitaire de Côte d'Ivoire. Cas des élèves et étudiants du district d'Abidjan. *European Scientific Journal,* 10.

Pronovost, G. (1994). Médias: éléments pour l'étude de la formation des usages sociaux. *Technologies de l'Information et Société (TIS*), 6(4) 377–400.

Proulx, S. (2005). Penser les usages des technologies de l'information et de la communication aujourd'hui: enjeux-modèles-tendances. In *Enjeux et usages des TIC: aspects sociaux et culturels*, Vieira, L. and Pinède, N. (eds). Presses universitaires de Bordeaux, Bordeaux, 1, 7–20.

Tano-Bian, A. (2015). La répression de la cybercriminalité dans les États de l'Union européenne et de l'Afrique de l'Ouest. PhD thesis, Université de René Descartes-Paris 5, Paris.

Tasso-Boni, F. (2014). La cybercriminalité au Bénin: une étude sociologique à partir des usages intelligents des technologies de l'information et de la communication. *Les Enjeux de l'information et de la communication*, 15(2B), 35–42.

Trépos, J.-Y. (1996). *La sociologie de l'expertise*. PUF, Paris.

Walter, J. (1995). *Les directeurs de communication. Les avatars d'un modèle professionnel*. L'Harmattan, Paris.

Winkin, Y. (2004). *Anthropologie de la communication. De la théorie au terrain*. De Boeck Supérieur, Paris/Louvain-la-Neuve.

# 7

# Money Transfers via Mobile Phones and Contributions to the Socio-Economic Well-Being of Populations in Africa

The contribution of information and communication technologies (ICTs) to the economic development in Africa has often been the subject of conflicting debates. Some support the idea that ICTs are an important lever for Africa's development; this is one of the main talking points surrounding the use of these devices. These talking points are mainly supported by development aid institutions and academics to name a few. Conversely, other actors take a more critical approach and believe that the potential of ICTs to contribute to Africa's development is often overestimated.

In evoking this debate, Alain François Loukou finds objectionable the idea that ICTs constitute an unproductive luxury in Africa in the light of traditional development priorities:

> "Considering moreover that investments in ICT cannot oppose those to be made in so-called priority areas, we believe instead that, in many cases, ICTs are now essential to the implementation of certain projects that have been identified as priorities." (Loukou 2012, p. 54)

As for Annie Chéneau Loquay (2010, p. 95), she believes that "no one denies the usefulness of the Internet and the telephone in Africa, even if the basic problems – water, energy and food supply – are still not resolved".

Chapter written by Évariste DAKOURÉ.

One of the most developed ICT devices in Africa, in terms of ownership and socio-economic impact, is the mobile phone. With regard to these impacts, Gaël Raballand makes the following observation:

> "Mobile phones not only reduce communication costs; they also have an impact on the financing of households in rural areas in some countries or even in terms of employment." (Raballand 2012, p. 739)

And to add:

> "Through SMS, mechanisms have been put in place in several sub-Saharan African countries to reduce waste and increase the selling price of agricultural goods." (Raballand 2012, p. 741)

In addition, concerning mobile phone activities in the informal sector in Côte d'Ivoire, for example, Dabié Désiré Axel Nassa (2011) indicates that the advent of the mobile phone:

> "has enabled the creation or even the production of more than three thousand direct jobs in Côte d'Ivoire. At the level of a municipality such as Adjamé, this tool has instead developed several related activities that are not recognized but permitted. Tens of thousands of indirect jobs are generated around these activities." (Nassa 2011 p. 5)

Continuing on the subject of job creation generated by the arrival of mobile phones in Africa, Henri Tcheng, Jean-Michel Huet and Mouna Romdhane (2010, p. 9) argue that "in Nigeria, where the telecoms sector, and in particular mobile telephony, is considered the main employer with 400,000 jobs created according to the ITU (International Telecommunication Union), mobile telephony promotes the development of an entrepreneurial spirit".

The financing of households of which Raballand (2012) is an example is not only done through the benefits of the abundant informal sector activities that mobile phones have generated in many African countries. It should be noted that these activities range from the sale of mobile phones and accessories to the sale of products via the use of the WhatsApp application as a channel for presenting products and making contact with potential customers.

The financing mentioned is also done through money transfers via mobile phones. These money transfers are an important sector in Africa. Referring to this importance, Alain François Loukou stated the following:

> "Money transfer is a highly developed activity in Africa because of the financial dependence of many people in rural areas on those in urban areas, and sometimes vice versa." (Loukou 2013, pp. 8–9)

In this chapter, we focus on the contribution of mobile phones to improving the socio-economic living conditions of African populations through an analysis of the impact of these transfers. Our study aims to show the extent to which mobile phones, particularly through fund transfers, can be considered as playing a significant role in improving socio-economic living conditions in Africa.

As mentioned above, money transfer activities in Africa are important for people, and it is for this reason that we have chosen for this study to focus on the contributions of money transfers via migrants' mobile phones. On the one hand, this choice is explained by the fact that the economic weight of migrants' money transfers is colossal. Indeed, this is a fundamental economic issue for the whole of Africa. "Another report published by the African Institute for Remittances (AIR) had already estimated that the amount of money transfers from the African diaspora to the continent reached $65 billion in 2017," according to Aboubacar Yacouba Barma (2018). The report states that this amount is more than double that of official development assistance to Africa, estimated at $29 billion in 2017. And, according to the report, the funds would benefit more than 120 million Africans through the operations of some 30 million African migrants.

On the other hand, we have decided to place particular emphasis on migrants' fund transfers via mobile phones, as the issue of international migration is a major political and socio-economic issue today.

This choice is also justified by the growing importance of money transfers via mobile phones in Africa. GSMA[1] (2018), referring to West Africa, states

1 GSMA represents the interests of mobile operators worldwide. It brings together nearly 800 operators and 300 companies belonging to the mobile ecosystem, including telephone and device manufacturers, software publishers, equipment suppliers, Internet service providers and companies in related sectors.

that "the number of mobile financial services subscribers registered in the region reached 104.5 million in 2017, an increase of 20.9% compared to last year". We highlight these elements, without minimizing other major economic issues related to ICT, and mobile phones in particular[2].

In this study, we support the idea that, given the scale of funds transferred from African migrants and the increase in transfer of funds by mobile phone, fund transfers from African migrants via mobile phones are significant in volume and have a high potential impact on the socio-economic living conditions of African populations.

After the methodological discussion, we will define the concepts required for the framework of our analysis. Then, we will discuss the continued growth in mobile phone penetration in Africa, which provides a basis for the development of mobile money transfer services. Then, we will present factors that contribute to the development of mobile money transfers. This analysis ends with a presentation of the results of our research, which explains how mobile money transfers from migrants contribute to improving the socio-economic living conditions of African populations.

## 7.1. Methodology

In this work, we have adopted a mainly theoretical approach, which has consisted in relying on the work of several authors who have already addressed this theme. We critique their work or use their approaches to support the ideas we put forward. Thus, we reviewed the literature on scientific articles and books, which was supplemented by a literature search based on two main lines of research: the use of reports and studies on the growth rate of mobile telephony in Africa and the increasing use of money transfer services; and the socio-economic impacts of migrants' money transfers via mobile phones.

2 In this regard, we wrote in 2014: "Cultural and communication industries – information and communication technologies, among others – now occupy an important place in the economies of rich and poor countries. The economic weight of this sector is driven in particular by the development of ICTs, and this gives rise to an abundant discursive production on the role of these devices in boosting the global economy" (Dakouré 2014).

The process consisted of drawing from the corpus of texts the speeches that allow us to support our hypothesis expressed previously. Quotation excerpts will be used for this purpose.

## 7.2. The notion of financial inclusion and mobile money

To define the notion of financial inclusion and to highlight that mobile phones promote financial inclusion, we refer to the work of Jean-Louis Arcand, Malek Garbouj and Nestor Morgandi. According to these authors:

> "In a broad sense, financial inclusion refers to any approach to providing affordable and non-traditional conventional financial services to vulnerable and marginalized groups that are partially or completely excluded from the formal financial system." (Arcand *et al.* 2013, p. 9)

As an illustration, they add that these groups may be low-income individuals, people with disabilities or migrants without legal status. These authors discuss the contributions of mobile phones to financial inclusion by indicating the following feature:

> "The notorious rise of mobile money and the resulting revolution in the financial industry have opened up new niches and avenues for financial inclusion for those excluded from the formal financial system." (Arcand *et al.* 2013, p. 2)

As for mobile money, they define it as money stored on a SIM card, which serves as an identifier but does not transit through a conventional bank account. In addition, we share Jean-Louis Arcand *et al.*'s point of view that one of the benefits of mobile financial inclusion is that it reduces transaction costs such as transport costs, time lost in travel and in queue before the transaction is concluded. According to Global Findex, 32% of individuals in sub-Saharan Africa are not "banked" because of the distance to a conventional financial institution (Arcand *et al.*, 2013 p. 11). In addition, mobile money also reduces the risks associated with holding liquidity, losing money or theft.

## 7.3. Mobile phone penetration rate in Africa and mobile money transfer

Before further discussing the contributions of mobile money transfers to the increase in financial inclusion in West Africa, let us first look at how the mobile terminal has established itself in this African space, including in rural areas.

Several factors have contributed to the growth of mobile phone access in Africa since the early 2000s. In this regard, Annie Chéneau-Loquay (2012, p. 85) expresses herself in these terms:

> "The 'miracle' of mobile telephony is linked to the arrival of cheap phones for less than $20, a dynamic second-hand market and, above all, the efforts made by manufacturers and operators who have adapted the product and services to users' practices."

Indeed:

> "From the early 2000s, they completely changed their model; rather than achieving a high margin on a small number of users, they preferred to achieve a low margin on a large number of subscribers, with a commercial policy that took into account low incomes."

In addition to making terminals accessible, industry actors have also made services accessible to low-income African populations. Telephone operators have extended network coverage and promoted technical innovations to facilitate their use.

Among other innovations, we can mention:

– "Prepayment, which concerns 96% of users, with cards and then electronic credit recharging systems, from phone to phone;

– the ability to transfer credit to another subscriber via text message with very small amounts of money;

– the offer of bonuses on an *ad hoc* basis, at certain times of the year;

– the considerable drop in the price of the SIM card ($5.12 in Senegal with the same amount in call credit in 2009, $2 in 2012)." (Chénau-Loquay 2012, p. 85)

As a result, in several capitals in West and Central Africa, new mobile phones (using two SIM cards) are quite easily available at very affordable prices. The use of dual SIMs allows users to play with operators' tariffs and adapt their consumption according to promotional offers. This is done by using one or another chip to make calls according to the network to which the desired correspondent's number belongs.

This penetration of mobile telephony concerns all areas, including poor or rural areas, which often lack social and technological infrastructure. Chéneau-Loquay maintains:

> "This relative levelling of the space of remote communication paradoxically allows the creation of low-cost uses in poor urban or peri-urban areas, under-equipped with fixed lines and power lines or even without them, as well as in rural areas." (Chéneau-Loquay 2012, p. 86)

Concerning African migrants in a precarious situation in Europe, they are not the only uses of mobile telephony services:

> "Aware of the benefits of even one mobile phone, migrants have not waited until the democratization of mobile teams and the (most recent) digital inclusion policies for them to benefit from it. Decoded mobile phone shared by several people, hunting down free communication zones, anything to speak a little longer with family, who have stayed in the country. But their use of ICT goes far beyond the simple communication function." (Gualtieri 2014b)

As we know, for family reasons, some African migrants, even in precarious situations, transfer funds to their relatives who have remained at home. Among them, some leave their African villages, leaving their wives and children, to try their luck in Europe. They will have access to mobile phones in their new host country.

Still with regard to the rapid development of mobile phones in Africa, Annie Chénau-Loquay thinks that it is because it is no longer the prerogative of the wealthy alone. Thus:

> "Small craftsmen and traders in the 'popular economy' sector, men and women of all walks of life, have adopted it because

> they have understood the value of the prepaid card subscription system (96% of users in Africa), and because they can easily access the multitude of services available to them near their workplaces or homes." (Chénau-Loquay 2012, p. 87)

These small craftsmen and traders are reluctant to use traditional banking services; as a result, access to mobile phones, and in particular to the mobile money transfer service, increases their financial inclusion.

According to the GSMA, at the end of 2016, West Africa had 172 million unique subscribers, representing 320 million mobile connections:

> "The region's penetration rate in terms of number of unique subscribers is now 49%, slightly higher than the 47% penetration rate recorded in sub-Saharan Africa as a whole." (GSMA 2017, p. 2)

According to the GSMA, mobile financial services, including mobile money services, continue to grow rapidly in West Africa as mentioned above.

In addition, the total value of transactions over the same period reached $5.3 billion. Furthermore, the report states that "there are now 57 mobile money services active in West Africa (compared to 51 in East Africa) and, in 2017, the region accounted for 34% of the total number of registered mobile money accounts in sub-Saharan Africa" (GSMA 2018, p. 23). Admittedly, this is a discourse accompanying an institutional actor that brings together the telecommunications sector. This could lead us to believe that this actor has every interest in supporting a positive vision of the contribution of m-banking and its development. However, this is an interesting indicator of the development of this type of service. The GSMA is not a provider of this service but an umbrella association.

## 7.4. Factors promoting the development of money transfers via mobile phones

The ease of use of the money transfer process via mobile phone by customers, some of whom have never used a bank card and ATM, has contributed to the success of this type of money transfer:

> "All you have to do is send the recipient a secret code by SMS, which remains active for a few hours. This ease of use probably explains the immediate success of the M-Pesa service. Shortly after its launch, the total value of money received by this method reached $100 million."[3] (McGovern 2011)

It is important to remember that in African countries there are more mobile phones than bank accounts. From this perspective, it is easy to understand that this market for money transfers via mobile phones represents enormous opportunities for various financial service players. These actors include post offices; money transfer companies such as Western Union and MoneyGram; banks; mobile phone operators; microfinance companies and several other small mobile money transfer service providers operating in the informal sector. It should be recalled that in this study, the main issue is the transfer of money via mobile phones.

After the launch of mobile money transfer in Kenya, there was a diversification of types of financial service delivery through this process. Thus, according to Anna McGovern (2011), another step has been to develop transactions that "eliminate or significantly reduce the role of cash in money transfer operations [...] We are already starting to see this when a person buys a prepaid communication package on an M-Pesa phone to send it to another user. There is no longer any need to pay cash [...] Thus, a person living abroad can send a prepaid communication package to the beneficiary, who can in turn exchange it directly for goods or services from participating retailers. Mobile financial services therefore involve international transfers, thus providing the possibility of direct transfers between migrants abroad and their relatives at home (including in rural areas). This practice of transferring mobile money to international destinations also exists in West Africa:

> "Airtel, MTN and Orange are among the mobile operators that have launched cross-border money transfers that have strengthened trade and commerce between the different countries in the region, including Burkina Faso, Côte d'Ivoire, Ghana, Mali and Senegal." (GSMA 2018, p. 24)

3 Available at: https://www.un.org/africarenewal/fr/magazine/december-2011/transferts-2%E2%80%99-Money-on-the-bank-second-phone.

The remainder of this chapter analyzes the impact of mobile money transfers from African migrants on local populations.

## 7.5. Presentation of the results

African migration flows mainly occur within the African continent. However, the financial flows linked to fund transfers from migrants living outside Africa are higher than those of migrants living in Africa.

### 7.5.1. *The issue of migrant fund transfers in Western Africa*

The Central Bank of West African States (BCEAO) published a report in 2013, from which it emerges:

> "that Africans emigrate mainly within the continent, with 64% of the total, ahead of OECD countries (25%). The phenomenon is even more pronounced at the sub-regional level, as indicated below… West Africa has the highest sub-regional migration rate, with 71 per cent of all migrants remaining in the area, compared to 24 per cent living outside Africa and 5 per cent in Central Africa." (BCEAO 2013, p. 10)

Moreover, the financial stakes of migration, particularly in terms of fund transfers from migrants, are higher for migrants living outside Africa. For most African countries with a strong migration tradition, funds from migrants living in European or American countries are significant. For example, the BCEAO's 2013 report states that resource flows to Mali come mainly from France (48.6%), Spain (10.5%), the United States (9.7%), the UEMOA[4] countries (8.7%) and the CEMAC[5] countries (11%). Concerning Senegal, another country with a strong migration tradition (to countries outside Africa), the same report states that "the European continent provides more than four fifths of the funds received (82.1%), of which 10.6% use

4 The West African Economic and Monetary Union includes eight countries: Burkina Faso, Benin, Côte d'Ivoire, Guinea Bissau, Mali, Niger, Senegal and Togo.

5 The Economic and Monetary Community of Central Africa consists of six countries: Cameroon, Central African Republic, Republic of Congo, Chad, Equatorial Guinea and Gabon.

informal channels. It is followed by Africa with 9.8% and the American continent for an estimated share of about 6.7% of the global total."

From this report, it emerges that three countries in the UEMOA region, including Burkina Faso, derive the largest share of fund transfers from migrants and their diasporas living in Africa. In this regard, the report states the following:

> "The majority of the resources received by Burkina Faso come from Africa (46.9%), especially the UEMOA countries (39.2%) and especially Côte d'Ivoire (30.9%). Europe is in second place with 32.0%, including the euro zone countries with 25.9%. The third largest provider of migrant workers' funds is the American continent with 16.3%." (BCEAO 2013)

The fact that Burkina Faso migrants' funds outside the country mainly come from West Africa is explained by the presence of a very strong diaspora in Côte d'Ivoire. This diaspora is estimated at more than 3 million Burkinabè today[6]. Even if Burkinabè living in Europe individually earn much more money than those living in Africa, the fact that there are many Burkinabè living in Côte d'Ivoire makes it possible to considerably increase the volume of migrant money transfers between that country and Burkina Faso.

The contribution of migrants' money transfers, mobile or non-mobile, is important in socio-economic terms for the relatives of migrants who have remained in their countries of origin. According to Naceur Bourenane:

> "Because of their importance for the recipient populations, migrants' money flows tend to remain stable and are not very sensitive to changes in the economic situation." (Bourenane 2011, p. 2)

Money sent from migrants is thus an important source of financing for the economies of African countries and their populations.

---

6 Mahamadou Zongo (2003, pp. 113–114) states that: "According to the 1998 General Population and Housing Census (RGPH), out of a total population of 15,366,672, Côte d'Ivoire had 3,954,550 foreigners, or 25.75% of the total population. Burkina Faso nationals, with a total of 2,238,548, constitute the largest foreign community: they represent 56.6% of foreigners and 14.56% of the total Ivorian population."

However, there are costs associated with these transfers. Some development cooperation actors are calling for these costs to be reduced so that they can participate more fully in co-development. On this subject, Bourenane indicates, in a report he conducted, the following fact:

> "The current trend towards stabilizing the costs of money transfers at a high level has led to national, bilateral and multilateral mobilization of States. This report therefore recommends solutions to reduce the costs of sending money and optimize their contribution to development." (Bourenane 2011, p. 3)

In the search for solutions that contribute to cost reduction of money transfers, it is also possible to see that with the development of ICTs, new money transfer processes and services are spreading and are helping to reduce these costs. This is the transfer of money by mobile phone. Indeed, in 2005, the GSMA officially launched a system of money transfers via mobile phones.

### 7.5.2. *Money transfers from African migrants via mobile phones and human development in Africa*

As a reminder, $65 billion was transferred to Africa by African migrants in 2017, according to a report by the African Institute for Remittances (AIR). Statistics on this type of transfer are also mentioned in an IFAD[7]report of November 2016, with growth estimates for 2020: "Money transfers to and within Africa will amount to nearly USD 80 billion in 2020" (Boon 2016, p. 10).

According to the study report, these money transfers affect more than 200 million people who are members of African migrant families, more than 40% of whom live in rural communities. Thus, money transfers from migrant workers represent a major support for many families and represent a significant economic weight in Africa; 200 million people, that represents a considerable number of Africans.

The increasing use of mobile phone money transfers is also contributing to these flows of money from African migrants from the African continent to their families. It should be noted that:

---

7 International Fund for Agricultural Development.

> "In a recent survey, GSM Association (GSMA) reviewed at least 29 cross-border mobile money transfer corridors between 19 African countries. This survey shows how this type of transaction across the continent has improved thanks to collaboration between telecommunications operators and money transfer companies." (Boon 2016, p. 12)

As an illustration, we mentioned that this type of transfer is offered by mobile phone operators in West Africa. In Burkina Faso, for example, an operator like Orange Burkina offers the possibility to transferring mobile money to six West African countries. On this subject, Orange Burkina specifies on its website:

> "You can send money from your Orange Money account to Togo (Moov), Guinea Bissau (Orange), Benin (MTN), Côte d'Ivoire (Orange), Mali (Orange) and Senegal (Orange)."[8]

Referring to the possibility for migrants to make money transfers via the telephone, Julia Gualtieri (2014a) suggested that since 2005 migrants can, by subscription, transfer money internationally via a simple call or by SMS, depending on the services. This pilot program appeared under the impetus of the GSMA and, according to Julia Gualtieri (2014a), it is "a change that is at least appreciated by migrants, many of whom have a telephone but few of whom have a bank account".

Continuing the discussion of money transfers via mobile phones, Annie Chéneau-Loquay points out the following fact:

> "The practice of transferring money by a simple phone call, which has existed for a long time, especially with migrants, and is based on trust and mutual assistance, has very quickly multiplied naturally via mobiles, even before operators have created specific procedures for these services." (Chéneau-Loquay 2012, p. 85)

One of the African countries where the mobile money transfer system has quickly become successful is Kenya. Several authors discuss the types of

8 Information on the Orange Burkina website. Available at: https://www.orange.bf/particuliers/1/5/envoyer-de-largent-28.html [Accessed May 28, 2019].

uses that African households receiving money from migrants make of this money. It appears that, apart from expenditure related to food, health, education and construction of buildings, which are very recurrent, some households invest in telecommunications and communication services to keep in touch with their relatives who have emigrated. We deliberately chose to highlight this aspect in order, among other things, to highlight the need felt by migrants' relatives to stay in touch and to have news of their relatives abroad. According to the BCEAO, in Burkina Faso:

> "66% of households receiving money from outside Africa have access to mobile phones, compared to 39% of those without any financial support. These households also have greater access to information through television (41% versus 9%) and the Internet (14% versus 2%)." (BCEAO 2013, p. 13)

The advent of mobile money is part of a logic that tends to reduce the cost of money transfers, whether for the benefit of migrants or not. In this sense, as Hans Boon estimates:

> "Transfers via dematerialized instruments (Internet, cards and mobile phones) will increase significantly in the medium term and will be offered at a lower cost than sending cash. At the same time, the number of rural recipients able to receive dematerialized money transfers is increasing." (Boon 2016, p. 45)

As rural African populations have increasing access to various mobile phone services, they will also benefit from this potential decline in money transfers. Thus, by assuming a decrease, migrants who send money to relatives in their home villages will be able to carry over the savings from sending costs to the very amount sent to their relatives, who in turn will be able to further improve their living conditions.

### 7.5.3. *Development of money transfer services in rural areas*

Money transfers from African migrants represent an opportunity for several actors in this financial sector. A survey conducted by BCEAO (2013, p. 8) on this type of transfer "reveals that a significant proportion of flows use informal channels, due in particular to high transaction costs and poor access to transfer services in rural areas". This suggests that there is a market for alternative players there, alongside large players such as banks and money transfer companies. To seize these opportunities, prerequisites are needed, which the BCEAO (2013) makes a case for, mentioning in particular:

– increased competition in the sector to encourage cost reduction;

– the development of rapid transfer service points in rural areas and the creation of new products to promote alternative operators such as post offices, savings and credit cooperatives or rural banks, which have extensive networks accessible to the poor and rural populations;

– the exploitation of the potential of *m-banking* for sending and receiving migrants' funds by mobile phone;

– intensifying studies on migration and workers' money transfers, for a better estimation of informal flows and economic reasons for these;

– the issuance of vouchers to capture diaspora placements, in order to increase migrants' opportunities to contribute to the financing of development projects in their countries of origin.

Since the birth of mobile money transfers in Africa, many studies have shown the continuous growth of this sector and pointed to the fact that there is a market for the transfer of money long distance and to rural areas on the continent:

> "In terms of long-distance monetary transactions, Naiditch (2009) shows that the advent of ICTs has completely disrupted the traditional way of transferring money in Africa, which remains a highly developed activity due to the financial dependence of many populations in rural and urban areas." (IRDES 2013, p. 55)

According to a report by the *Institut de recherche pour le développement économique et social* (IRDES) of Senegal, this mode of transfer is developing between African cities because it is considered simple, reliable and fast:

> "Indeed, in the moments following the deposit of the money via an approved agency, and provided that he is informed by mobile phone, the recipient can immediately withdraw the money transferred to him by going to a local agency of the various companies operating in this field." (IRDES 2013, p. 55)

In the South African case, Anna McGovern (2011) reports that:

> "Mobile phone transactions can allow banks to access rural markets without having to open new branches. By May 2011, nine months after its launch, the Vodacom-Nedbank partnership had 140,000 subscribers in South Africa and had established more than three thousand M-Pesa branches and two thousand ATMs across the country."[9] (McGovern 2011)

The limitation to be mentioned in these cases is that sometimes rural areas are underserved and often some small mobile money transfer service providers, who operate in the informal sector, have very small amounts of money.

Nevertheless, solutions exist to address the low access to financial services in rural Africa. In this sense, alliances are being formed, involving mobile phone operators, banks and microfinance actors (who generally have a wider network, including rural areas) to enable people in these rural areas to have access to a mobile money transfer service. There are examples of this in Ghana, to cite just one example:

> "The mobile money transfer network set up through partnerships with four mobile phone companies in Ghana has enabled us to expand our services to customers across the country." (McGovern 2011)

In some cases in Burkina Faso, even when the mobile phone is not directly used for money transfer, it is one of the tools used in the money transfer process. On this subject, Sylvestre Ouédraogo (2008, p. 21), after an interview with the head of the payment and transfer department of the Caisses populaires du Burkina network, states that "from this interview it appears that the mobile phone is mainly used until then for the confirmation and verification of transactions of transfers, the main tool used being the fax machine". As an illustration, Ouédraogo mentions this characteristic:

> "For a transfer between a branch in Ouagadougou and a branch in Bobo, the customer's transfer orders and account numbers

9 Available at: https://www.un.org/africarenewal/fr/magazine/december-2011/transferts-2%E2%80%99-Money-on-the-bank-second-phone.

> are faxed and it is after that that that the mobile phone intervenes to confirm. The laptop is mainly used instead of the fixed-line because the agents are all in the fleet system." (Ouédraogo 2008, p. 21)

The fleet is a subscription to several grouped mobile phone numbers, at a preferential price, that an institution can subscribe to from a mobile phone operator to allow each employee with a SIM card in the group subscription to call another employee with a SIM card in the same subscription for free. The price of each communication is paid monthly by the subscribing organization for its employees. In this case, it is a tool to facilitate communication, including in a mobile situation, between employees of an organization.

### 7.5.4. *Mobile money transfer and inclusion in Africa*

We mentioned how mobile financial services continue to grow in West Africa. According to the GSMA, this growth is due to an increasingly favorable environment for this type of service, including:

> "the rapid expansion of agent networks and a favorable regulatory environment. In West Africa, there are 13 times more active mobile financial services agents than bank branches and ATMs, and 13 out of 15 countries have enabling regulations for their mobile money services." (GSMA 2018, p. 23)

Thus, the increase in active accounts for mobile money transfers contributes significantly to financial inclusion in this part of Africa, where, as mentioned above, few people have access to traditional financial services. The GSMA states this:

> "With more than 400 million registered users, mobile money facilitates access to financial services, many of which contribute to strengthening the resilience of the poor by reducing their vulnerability to disasters or economic, social or environmental shocks." (GSMA 2017, p. 13)

Thus, mobile money transfers from migrants also contribute to the resilience of poor rural and urban populations, who would be in more difficult living conditions if they did not receive these transfers.

It is important to also mention the following characteristic:

> "For many marginalized groups around the world, mobile telephony is the first step towards a connected society. Mobile telephony plays a key transformational role in providing a formal identity and access to various basic services, including financial services." (GSMA 2017, p. 15)

Among these important transformational roles of the mobile phone is the fact that, by receiving money transfers from African migrants, it can significantly impact the socio-economic development of rural areas and even entire countries.

## 7.6. Conclusion

This chapter provided an opportunity to take part in the debate on the contributions of ICTs, and particularly mobile phones, to improving the socio-economic living conditions of African populations. By following a logic where we do not deny the existence of multiple priorities in social areas that require large investments from governments, we do not agree with the theses of those who believe that ICTs are a luxury for Africans who should focus mainly on other priority areas. It is for this reason that this study has focused on showing how ICTs, through mobile money transfers, especially those of African migrants living in or outside Africa, contribute to improving the living conditions of African communities.

The chapter showed that the field of mobile telephony is a major socio-economic sector in Africa that creates jobs, generates economic activity, especially in the informal sector, and boosts the transfer of money to the continent, by providing the possibility of transferring funds at a lower cost. It shows that this opportunity is an aid for African migrants, many of whom transfer considerable amounts of money to their relatives in their country of origin and often to rural areas.

Thus, we show that migrants' mobile money contributions are a significant contribution to improving the socio-economic living conditions of millions of African populations. This is all the more so as there are opportunities in terms of growth in migrant money transfers via mobile phones. Indeed, money transfers from African migrants generally constitute a large financial windfall. These transfers far exceed Official Development Assistance (ODA) in volume.

They are also higher than Foreign Direct Investment (FDI) in Africa, so much so that we could say that African migrants are the main funders or investors in Africa. This provides a better understanding of the economic challenges associated with international migration and the difficulties in regulating irregular migration: these migrants who have arrived in a European or American host country represent potential financial windfalls.

This chapter has also shown how the development of m-banking contributes to financial inclusion in Africa, by providing an increasing number of Africans with access to financial services, which are not just the transfer of funds, although it is this aspect that interests us mainly here. These m-banking activities concern other widely used services such as payment via mobile money, for example. This type of service is becoming increasingly widespread, even in poor or rural areas where mobile phones can provide new services. It would therefore be interesting, at a time when connectivity is increasing in Africa, to analyze the impact of access to mobile financial services on trade and communications between economic actors in rural Africa.

## 7.7. References

Arcand, J.-L., Garbouj, M., and Morgandi, N. (2013). Transfert de fonds et services financiers sur mobile: les modèles d'affaires pour les Postes. *Global Migration Research Paper*, 6.

Barma, A.Y. (2018). Transferts d'argent: les migrants africains toujours généreux [Online]. Available at: https://afrique.latribune.fr/economie/conjoncture/2018-04-24/transferts-d-argent-les-migrants-africains-toujours-genereux-776357.html.

BCEAO (2013). Synthèse des résultats des enquêtes sur les envois de fonds des travailleurs migrants dans les pays de l'Uemoa [Online]. Available at: https://www.bceao.int/sites/default/files/2017-12/synthese_des_resultats_des_enquetes_sur_les_envois_de_fonds_des_travailleurs_migrants_dans_les_pays_de_l_uemoa.pdf.

Boon, H. (2016). Transferts d'argent et bureaux de poste en Afrique. Répondre aux besoins des migrants et de leurs familles en milieu rural [Online]. Available at: https://www.ifad.org/documents/38714170/40194630/PostalOffices ReportWEB_f.pdf/3199a789-1be3-43d4-9503-988d05b3b947.

Bounie, D., Diminescu, D., and François, A. (2010). Une analyse socio-économique des transferts d'argent des migrants par téléphone. *Réseaux*, 1(159), 91–109.

Bourenane, N., Bourjij, S., and Lhériau, L. (2011). Réduire les coûts des transferts d'argent des migrants et optimiser leur impact sur le développement. Outils et produits financiers pour le Maghreb et la Zone franc [Online]. Available at: https://www.afdb.org/fileadmin/uploads/afdb/Documents/Publications/201112Transferts_argent_migrants_resume_fr-updated.pdf.

Bourven, M. (2008). Comment mobiliser l'épargne des migrants en faveur du codéveloppement. Rapport du Conseil économique et social (France) [Online]. Available at: https://www.lecese.fr/sites/default/files/pdf/Avis/2008/2008_14_moniquebourven.pdf.

Chéneau-Loquay, A. (2010). L'Afrique au seuil de la révolution des télécommunications. Les grandes tendances de la diffusion des TIC. *Afrique Contemporaine*, 2(234), 93–112.

Chéneau-Loquay, A. (2012). La téléphonie mobile dans les villes africaines. Une adaptation réussie au contexte local [Online]. *L'espace géographique*, 1(41), 82–93. Available at: https://www.cairn.info/revue-espace-geographique-2012-1-page-82.htm.

Dabié, D.A.N. (2011). Contribution de la téléphonie mobile à la dynamisation du commerce informel dans la commune d'Adjamé à Abidjan en Côte d'Ivoire [Online]. 1–151. Available at: https://halshs.archives-ouvertes.fr/halshs-00655619/document.

Dieng, S.A. (2000). Épargne, crédit et migration: le comportement financier des migrants maliens et sénégalais en France [Online]. PhD thesis, Université de Lyon 2, Lyon. Available at: http://theses.univ-lyon2.fr/documents/lyon2/2000/dieng_sa/download.

Dieng, S.A. (2008). Déterminants, caractéristiques et enjeux de la migration sénégalaise [Online]. *Revue Asylon(s)*, 3, March. Available at: http://www.reseau-terra.eu/article709.html.

Gemenne, F. (2015). Migrations internationales: un enjeu Nord-Sud? [Online]. *Alternatives Sud*, 22, 1. Available at: https://www.cetri.be/IMG/pdf/as_migrations_pdffinal_2015_-_1.pdf.

Gualtieri, J. (2014a). La banque portable, l'outil du migrant autonome [Online]. Available at: https://digital-society-forum.orange.com/fr/les-forums/138-la_banque_portable_lroutil_du_migrant_autonome.

Gualtieri, J. (2014b). De l'intégration sociale à l'inclusion numérique [Online]. Available at: https://digital-society-forum.orange.com/fr/les-forums/141-de_lintegration_sociale_a_linclusion_numerique.

Institut de recherche pour le développement économique et social (IRDES) (2013). Étude sur les transferts de fonds des migrants sénégalais [Online]. International investigation report. Dakar. Available at: http://www.dmc.finances.gouv.sn/fondsmigrants_senegalais.pdf.

Loukou, A.F. (2011-2012). Les TIC au service du développement en Afrique [Online]. *tic&société*, 5(2–3). Available at: https://journals.openedition.org/ticetsociete/1047 [Accessed 30 April 2019].

Loukou, A.F. (2013). Les techniques d'information et de communication (TIC) et l'évolution de l'économie africaine: vers une hybridation des activités. *Les enjeux de l'information et de la communication*, 14(1), 103–116.

McGovern, A. (2011). Transferts d'argent: le téléphone portable au secours des banques [Online]. Available at: https://www.un.org/africarenewal/fr/magazine/december-2011/transferts-d%E2%80%99argent-le-t%C3%A9l%C3%A9phone-portable-au-secours-des-banques.

Ndiaye, M. and Robin, N. (2010). Les migrations internationales en Afrique de l'Ouest: une dynamique de régionalisation articulée à la mondialisation [Online]. *Dossier: Les migrations subsahariennes*. Available at: http://horizon.documentation.ird.fr/exl-doc/pleins_textes/divers18-07/010057435.pdf.

Ouedraogo, S. (2008). Rapport provisoire de l'étude sur les transferts et l'usage des terminaux mobiles au Burkina Faso [Online]. Association Yam Pukri, Burkina Faso. Télécom ParisTech. Available at: http://www.burkina-ntic.net/ressources/assets/docP/Document_N081.pdf.

Raballand, G. (2012). Le téléphone mobile a-t-il créé une révolution en Afrique?. *S.E.R Études*, 6(416), 739–748.

Tcheng, H., Huet, J.-M., and Romdhane, M. (2010). Les enjeux financiers de l'explosion des télécoms en Afrique subsaharienne. *IFRI Programmes Afrique subsaharienne*.

Zongo, M. (2003). La diaspora burkinabè en Côte d'Ivoire. Trajectoire historique, recomposition des dynamiques migratoires et rapport avec le pays d'origine. *Politique africaine*, 2(90), 113–126.

# 8

# Geographic Concentration of Monetary Activities and Spatial Dispersion induced by Digital Technology in Côte d'Ivoire

For many decades, monetary activities in Côte d'Ivoire were mainly organized, from a geographical point of view, around a centralized type model. In this model, only large urban centers benefited from financial institutions, while a very large part of the national territory lacked monetary services. The economic and commercial strategies developed by the players in the field undoubtedly explain this concentration (World Bank 2003).

However, it appears that the technological devices formerly used in the trade are not foreign to this form of geographical distribution (Loukou 2013b; Loukou 2016). Telecommunications techniques were not widespread, and computer and electronic systems for remote networking of activities were very underdeveloped in the country. Towards the end of the 20th Century, the rapid development and widespread use of information and communication technologies would lead to an unprecedented upheaval in banking activities. Following Clayton Christensen (1997), it was then commonly accepted that a technological disruption or disruptive innovation characterized the profound transformations brought about by digital technology in economic activities. Christensen's pioneering work (Christensen 1997; Christensen 2003), from which he introduced the concept of disruptive technology, provides an original insight into how disruptive

Chapter written by Alain François LOUKOU.

innovation[1] generally involves new business models as well as technological innovation. Thus, in the field of financial transactions in Côte d'Ivoire, there is a genuine example of digital disruption leading to new services, such as mobile money, and a new organization of work in the sector. These innovations have transformed the relationships, behaviors and feelings of Ivorian consumers towards the national financial system (particularly in terms of preference, frequency of transactions and assurance of speed and geographical proximity). Beyond that, the breadth of the scope of digital money activities has generated a new and completely innovative geography of national money activities. Henceforth, this geography, governed by electronic flows, is characterized by a fragmented spatial distribution of monetary transaction activities according to a networked mode of operation of the territories, and no longer by a simple agglomeration or a juxtaposition of a few areas. By their intrinsic ability to allow the spatial dispersion of activities, ICTs are now proving to be real tools for decentralizing service activities in the territories.

The work is summarized as follows: after describing the theoretical and spatial frameworks and the way in which data are mobilized, the study will focus on analyzing, in terms of results, four key points:

– the classic model of concentration of monetary activities in Côte d'Ivoire;

– the model of spatial decentralization induced by the vitality of digital technology;

– the interpretation of contemporary geographical space through the dynamics of digital technology;

– the financial inclusion of a growing proportion of the population through digital technology. The analysis of the new economic model and the mechanisms of territorial logic by ICTs will then make it possible to perceive the link between digital technology and the decentralization of activities in the territories of Côte d'Ivoire.

Finally, the discussion will highlight the various barriers to digital transactions and the limitations of the methodology adopted.

1 The concepts of "technological disruption" and "digital disruption" are also used to refer to the profound transformations brought about by digital technology in economic activities.

## 8.1. Theoretical framework of the study: a fundamental disruption to traditional economic models, linked to the digital revolution

The digital revolution, following the rapid development of the global information and communication technology (ICT) sector, has led to a fundamental disruption to historical economic models (Christensen 1997; Christensen 2003; Sadin, 2013). This disruption has occurred in two ways: by improving production processes in the business world, and thus by modifying the value chain of companies as well as by developing new strategies, designing new business models and introducing new services such as payments/cash transfers via mobile phones. These radical changes in the technical and economic order have led to sectoral reorganizations, new habits among populations and the migration of certain historical sectors to new sectors. These changes are referred to by the new concepts of digital disruption or disruptive innovation. These concepts suggest a considerable disruption of existing business models, caused by any new technology, particularly of a digital nature. Globally, the most typical examples at the moment are Netflix, AirBnB, Uber and Spotify. These new companies are respectively active in providing services such as online television series streaming , online rental and reservation of accommodation, operation of mobile applications to connect users with drivers offering transport services and music streaming. The digital-intensive business model of these companies is definitely revolutionary[2]: in just a few years, they have succeeded in overtaking the old ones with their innovative concept, constantly putting them under pressure. Disruptive innovation therefore implies a new business model dictated by digital technological innovation.

In Côte d'Ivoire, and more broadly in Africa, in the specific field of monetary transactions, we are witnessing, since the social and geographical generalization of digital tools, an unprecedented example of innovative disruption that is the source of new services, and above all of a new geography of these activities. Indeed, we have moved from the primitive model characterized by the geographical concentration of monetary transaction activities in a few large urban centers to a model of broad spatial decentralization of these activities in many territories (both urban and rural).

---

2 Building on the services already created by Web giants, known by the acronym GAFA (Google, Apple, Facebook, Amazon), Netflix, AirBnB, Tesla and Uber (NATU) are a new generation of companies that are fully exploiting digitalization to develop new disruptive business models that make the rules of their core market obsolete.

In terms of geographical expectation, accessibility and mobility, Ivorian consumers are having new transactional experiences that are changing their habits, behaviors and relationships with the national and even international financial system.

## 8.2. Description of the spatial framework of the study

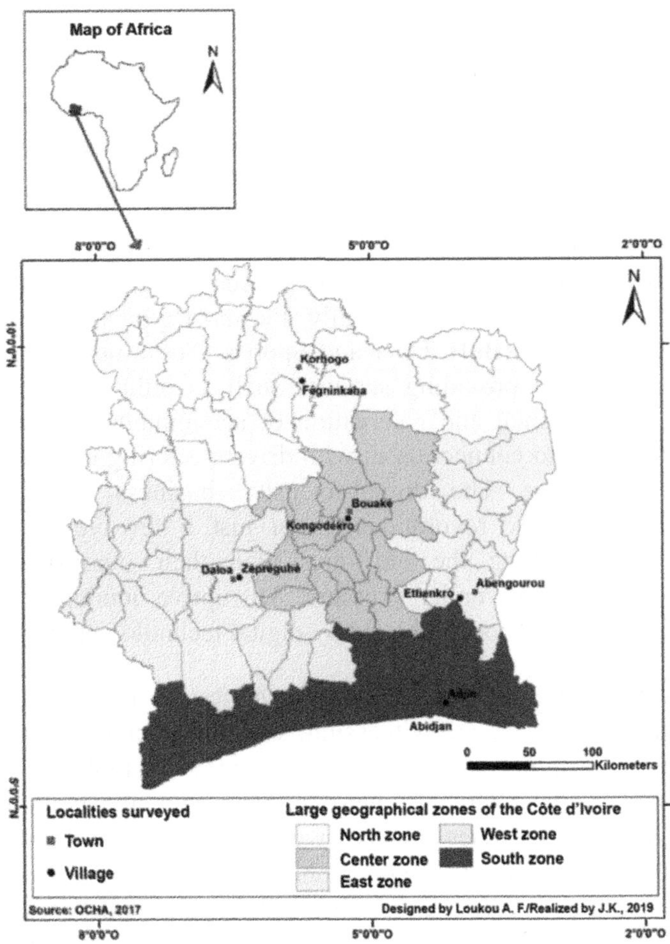


**Figure 8.1.** *Location map of Côte d'Ivoire in Africa and survey areas. For a color version of the figures in this book, see www.iste.co.uk/atchoua/digital1.zip*

The geographical framework considered for this study is Côte d'Ivoire where, like most other countries on the continent, access to financial services is considered, via ICT, to be radically disrupting the economy and the transactional habits of populations. The ICT Development Index (IDI), which compares the performance of ICTs between countries around the world, was 3.14 in 2018. As a result, Côte d'Ivoire ranked 131st out of 175 states (UIT 2017). Nevertheless, it should be noted that the digital sector is developing relatively quickly in the country thanks to mobile access technologies, which makes it possible to increasingly structure the development of monetary activities around digital technology, both in rural and urban areas. Based on a population of approximately 22 million in 2014 (RGPH 2014), the country's traditional banking rate was estimated at 21.7% in 2018 according to the Central Bank of West African States (BCEAO). As for the ratio of adults with a mobile account in Côte d'Ivoire, it was 49.45% as of June 30, 2018 according to the *Autorité de régulation des télécommunications en Côte d'Ivoire* (ARTCI). These various statistics relating to the study space are interesting in view of the reflections that will follow in the analysis of the results and their discussion. Figure 8.1 shows the location of Côte d'Ivoire in Africa and the areas of the country where the surveys were conducted.

## 8.3. Methodology

In their methodological approaches (in terms of surveys), the numerous empirical studies already undertaken on digital financial services have too often focused on the technical and socio-economic dimensions and neglected the spatial factors (location) of the phenomenon. Also, as part of this study, in order to better assess the spatiality of monetary transaction activities at the national level, our methodology consisted, in addition to observations, of conducting field surveys in five geographical regions of the country (north, south, east, west and center). Within each of these regions, we focused on a large city and a large village (Figure 8.1), in order to have a clear idea of the threshold of monetary activity in these two types of environments. This approach has the advantage of knowing whether the activity is polarized or not. It also makes it possible to appreciate the contrast between the city and the countryside in the distribution of services. On this basis, three aspects of the economic model shift were taken into consideration in data mobilization: the territorial, commercial and technological aspects. The target population was mainly users of these digital financial services. Depending on the number of branches and their visit frequency by users of the survey areas, we randomly interviewed 120 people, 75 of whom were in urban areas and 45 in rural areas. The objective was to capture their level of satisfaction with this innovative service. The level of satisfaction is reflected in the parameters of geographical accessibility to agencies' payments/transfers, speed of operations, ease or complexity of procedures and quality of operators' communication networks.

## 8.4. Results

### 8.4.1. *The classic model of concentration of monetary activities in Côte d'Ivoire*

Before the advent and generalization of electronic money transfer and payment systems in Côte d'Ivoire, including such modalities by mobile phone, monetary transactions between remote sites were characterized by a business model that was not very widespread, or even polarized, like that of traditional banks. Indeed, they were essentially carried out from a few central locations. The geography of these activities was directly controlled by the main means and modes of transfer then used: post offices, certain interurban transport companies and "person-to-person". As a corollary to this old model, when an individual wanted to send money to a third party (a relative or acquaintance) residing in any part of the country, three main physical solutions were available to him or her: either he or she made a money order, or he or she used the informal services of an inter-city transport company serving the area in question, or he or she sought the goodwill of an acquaintance who was traveling to the area in a timely manner. Whichever option was chosen, the money only reached the recipient as soon as possible, after several hours or even several days, since it was, so to speak, literally conveyed physically (except in the case of the money order). While post offices were represented in most cities in Côte d'Ivoire, their activities were limited to cities, depriving the populations of the surrounding rural territories of services such as postal money orders. As for the financial transactions carried out by the interurban transport companies, their geographical area of influence was determined by the topology of their services, which were also highly centralized: in general, from Abidjan to the inland cities and in the opposite direction or at best between nearby cities. Finally, the "person-to-person" mode was more than random, a matter of simple providence: the time of sending money should absolutely coincide with the departure of a trustworthy person to the locality concerned (unless a special envoy is dispatched in case of extreme emergency).

By their nature, structure and modality of operation, these three systems were naturally destined to be spatially undiversified. They did not promote a homogeneous geographical dispersion of transaction activities throughout the national territory. Also, the journey that people in rural areas generally had to undertake in order to carry out their financial operations in urban areas, *de facto* conferred on the latter a status of central places around which peripheral areas gravitated (Figure 8.2). Fortunately, the advent of ICTs would quickly disrupt the monetary ecosystem and make these models obsolete. Thus, from the traditional model of geographical concentration of monetary activities in the main cities, we have moved on to a model characterized by a spatial fragmentation of these activities throughout the national territory thanks to electronic networks and transfer systems.

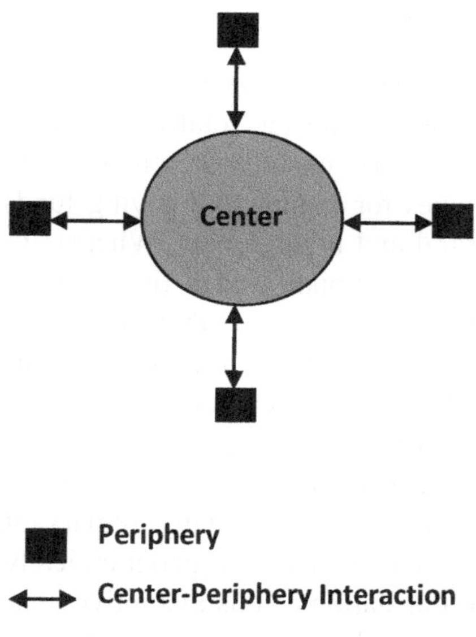


**Figure 8.2.** *Centralized Service System Model of payment/transfer around a geographical point. For a color version of the figures in this book, see www.iste.co.uk/atchoua/digital1.zip*

The main disadvantages of the centralized system are as follows:

– few centers likely to be linked to various peripheries;

– slowness in the processing, management and execution of operations;

– loss of time;

– reduced system efficiency;

– minimum user satisfaction.

### 8.4.2. *The model of decentralization of monetary transaction activities, driven by the vitality of digital and electronic technology*

By their intrinsic characteristics (possibility of more widespread spatial diffusion of phenomena) and their unprecedented vitality, digital technologies allow a new form of deployment of the productive force (means and populations) on the territories. As a result, they are now real tools for decentralizing service activities.

Today, the digital world of computers, mobile cellular phones and massive data transmission systems has totally revolutionized the way money is sent to Côte d'Ivoire. It has enabled the creation of a particularly dense network of small agencies for issuing and paying funds, spread throughout the country (in both rural and urban areas). Whereas in the past, following the centralization of the traditional model, financial transactions were only carried out from a few central points to other localities, it is now possible to carry them out in various localities to the great delight of the populations. This digital disruption has thus created a new spatial structure for monetary activities in Côte d'Ivoire. Indeed, regardless of the ease, reliability and extreme speed that characterize this innovative procedure, and which have made the business evolve, it is especially remarkable to note that this technological disruption has generated a geographically fragmented model. It is mainly the spatial translation of this disruptive innovation characterized by an unprecedented geographical dispersion of monetary activities in Côte d'Ivoire, which has made it an enormous success with Ivorian populations.

Almost anywhere, people are assured that they can send or receive money quickly if they need it. A good majority of them (about 50%, according to ARTCI statistics) no longer have to travel very long distances for these services, which are all in all basic, even universal. Several types of companies – specialized companies (Western Union, Money Gram, etc.), banks, mobile cellular operators, post office, transport companies, etc. – operate in this commercial and technological niche to serve users.

Analysis shows that the migration from the old model to the new model is driven both by operational necessity and by a need for circumstantial adaptation driven by technological developments in the financial sector. This system is undoubtedly a successful example of the use of digital tools to serve citizens throughout the territory, while at the same time constituting a significant factor in spatial planning, reducing the digital divide and socio-economic development (Loukou 2013b).

In the absence of sufficient data to map the spatial location of these activities, this deficiency can be validly remedied by a combination of geographical modeling: Christaller's theory of central places and Brunet's chorematics.

These two systems allow spatial phenomena to be represented in their complexity using graphical models, as shown in Figures 8.2 and 8.3.

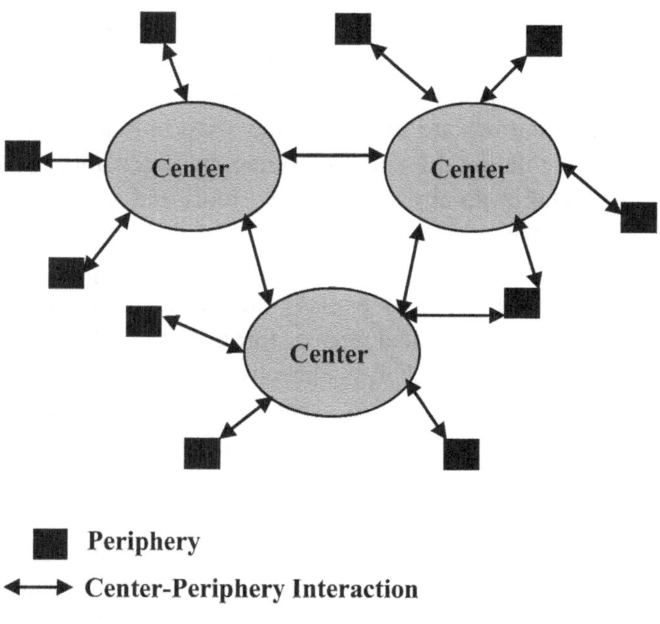


**Figure 8.3.** *Decentralized system model digital payment/transfer services. For a color version of the figures in this book, see www.iste.co.uk/atchoua/digital1.zip*

The main advantages of the decentralized system are as follows:

– multitude of centers and peripheries;

– speed of processing, management and execution of operations;

– time saving;

– better system efficiency;

– increased user satisfaction.

## 8.5. The interpretation of geographical space through digital services

Now highly integrated into various human activities, digital tools are now powerful means of interpreting space through the classic epistemological questions of geography: where? When? When? How? Why?? What is the result of this or that action? For illustrative purposes, and in relation to our analysis, here are some examples of questions raised in a hierarchical logic (from the simplest to the most complex): where can we find the most digital payment services? When did this type of service start? How are these activities implemented in the geographical area? Why are there more mobile payment services in one location than in another? What is the consequence of such spatial distribution on socio-economic development? If spatial regularities or irregularities are found, do they exist only at localized scales or do they occur at the national level? Of course, by extension, these questions can be asked in relation to any of the digital tools and services that structure the contemporary geographic landscape (telephone, Internet, radio, television, etc.). The objective answers to these questions make it possible to provide reliable and operational keys to explain the mechanisms and logic of the functioning of the geographical space through the electronic monetary transaction activities for which it is the physical support. Still in terms of interpretation, we also observe that from the simple traditional agglomeration of a few geographical places of monetary activity, we have achieved a functioning of financial territories in network mode thanks to digital technologies. Thus, through the many activities it creates, digital technology is constantly generating new spatial structures and new topologies that today allow the financial inclusion of many Ivorian populations, who until now have remained on the sidelines of the traditional financial circuit.

## 8.6. The financial inclusion of a significant proportion of the population through digital technology

Financial inclusion is generally defined as access to financial services, at a lower cost, for populations excluded from the banking system. These populations include both urban and rural populations. Literally, financial inclusion therefore aims to include marginalized populations in the financial system. It can be carried out through various kinds of mechanisms and modalities (technological innovations, decentralization of existing services, creation of new services, provision of facilities).

With the democratization of digital technologies (especially mobile cellular telephony) to all social strata and their widespread diffusion in various territories, digital financial services quickly emerged as the ideal solution for the financial inclusion of populations in Africa. Digital technologies are undoubtedly a means of implementing effective solutions flexibly and quickly and offering appropriate products to different segments of the population. More than 90% of the individuals surveyed in our surveys said that they now prefer mobile payment systems because of a combination of factors that characterize them: their geographical accessibility, the ease and speed of operations, the time savings they offer in operations and the relatively low costs of the transactions they carry out. They rarely find these advantages with conventional banks, which they consider rather cumbersome, complicated and, above all, slow in their counter operations. Some also find that ATMs are often card "swallowers". Others even feel that traditional banks paradoxically exclude part of the population through their policies, which they consider to be generally unenticing. In contrast, they believe that the practicality of digital banks is a guarantee of financial inclusion for the poor. Therefore, even if they have a basic current account in the traditional financial system, they find it useful to have one or even two mobile money accounts.

The World Bank's 2018 report on the economic situation in Côte d'Ivoire provides interesting statistics and analysis on the power of financial inclusion of populations through digital transaction systems, including mobile payment. The report notes that the monetization of the economy has increased due to the relatively rapid growth in the number of Ivorians with a bank account: 41% in 2017 compared to 34% in 2014. The report points out that this rapid progress is mainly a reflection of the explosion of digital accounts with mobile phone operators, while traditional bank accounts remain modest.

In fact, the report clearly shows that Ivorian savers are increasingly abandoning banks in favor of mobile phone operators. For its part, the Telecommunications Regulatory Authority of Côte d'Ivoire (ARTCI 20108b) estimated (based on more recent figures) that with a mobile phone penetration rate of 131.6% as of June 30, 2018, the country has a mobile money account rate of 49.45%. In terms of turnover, the Ivorian Ministry of the Digital Economy notes that in 2018, 17 billion CFA francs of financial transactions were carried out on a daily basis through mobile payments. Considering that this activity operates every day of the week, a basic calculation gives an annual transaction of 6,205 billion CFA francs.

## 8.7. Discussion

### 8.7.1. *Limits of geographical coverage of mobile transmission networks and technical difficulties of inclusion*

The operationalization of mobile payment is primarily conditioned by telecommunications networks and systems. In fact, from a technical point of view, the procedure is based on the convergence of electronic, computer and data transmission systems commonly known as ICT. However, in many rural areas, the permanence of so-called gray and white[3] areas does not always facilitate transactions, to the extent that the populations of these areas frequently have to travel to urban areas to conduct their operations. The geographical reading that results from this situation is that, in general, the distribution of human activities in the geographical space follows two main types of trends: homogeneity or heterogeneity. In this case, the distribution of mobile payment services in Côte d'Ivoire is not exempt from this duality, which includes different variants such as polarization, centrality, peripheral areas and equal or unequal distribution. Admittedly, the remarkable development of digital technology has led to a very significant decentralization of monetary activities in Côte d'Ivoire. However, in many rural areas, the persistence of gray and white areas, as mentioned above, suggests a local distribution that is still uneven or even polarized in some areas. However, all in all, it can be said that the geographical distribution of

3 In telecommunications jargon, the term "gray areas" refers to geographical areas where access to the Internet or mobile phones is characterized by a much lower signal level than in other areas. As for the "white areas", it refers to geographical areas that are not served by an electronic communication service (mobile telephony and Internet) at all. In either case, it boils down to the digital divide.

services tends to be increasingly homogeneous and that the advent and democratization of mobile payment, which has led to a vast network of small payment/transfer agencies, has also considerably changed people's relationships with the financial system.

### 8.7.2. *The mobile payment solution: advantages and disadvantages*

Despite the advantages it offers (geographical accessibility of payment points, time saving, acceleration of monetary circulation, etc.), the mobile payment solution unfortunately comes with certain disadvantages and abuse. This mainly concerns cybercrime consisting of various scams and fraud, carried out via mobile phones. Some agents involved in the activity are strongly suspected of being the main perpetrators of these offences. Another disadvantage consists of so-called "transfer to a third party" or "remote code confirmation" procedures for which the use of secret codes is necessary to validate an operation. Technically, these practices breach the principle of confidentiality of the secret code that characterizes any electronic financial procedure. In general, it is illiterate users who are unable to make withdrawals themselves who are forced to use the services of a third party. Similarly, some populations residing in rural areas without mobile withdrawal/transfer agencies also use these processes at the risk of being robbed by dishonest agents. In addition, the World Bank report, cited above, notes that while digital services via mobile tools facilitate payments and money transfers, they do not allow credit, one of the main functions of any financial system.

## 8.8. Limitations of the study from a methodological point of view

Based in part on information from customers, money payment agencies, mobile operators and, secondarily, local authorities, the methodology used loses objectivity in a certain sense, given that these different actors do not always provide the quality of information required. In fact, in this study, the field observation criterion was more relevant than any other criterion, especially when you are yourself, beyond the status of researcher, a direct witness to the revolution taking place, and occasionally a user of these services. Gathering the ability to compare recent past monetary transaction activities with the current situation was decisive. However, while it appears overall that the description and analysis are in line with the observational evidence in the field, there are still some gray areas of understanding. Indeed, in order to grasp the full national territorial and societal implications of the digital payment revolution, it would have been interesting to have more refined and broader data, which was not possible due in particular to the mistrust of the commercial actors operating in the sector. This data gap has led to the use of a combination of graphical modeling (see Figures 8.2 and 8.3) to address the inability to produce reliable geographical maps to spatially represent digital transaction services. These are the "central places" and "chorematic" models. However, despite their proven usefulness, these graphical models have limitations: existence of accentuated gradients between the centers and the peripheries, lack of precise geographical location of the phenomena studied, etc. Under these conditions, it is inherently difficult to know whether the model is geographically uniform, in other words, whether it contains local irregularities here and there. At most, with the rapid pace of change in activity, correlated with the rapid development of digital technology and the growing need of populations for financial inclusion, it is possible to make favorable predictions regarding an increasingly wide and dense geographical distribution of these services throughout the national territory.

## 8.9. Conclusion

Gradually, work on digital technology in society and in the territories is beginning to stabilize in terms of the general picture of theoretical and empirical thinking. This study aimed to contribute to strengthening this scientific basis. While digital technology is at the heart of scientific concerns as a research subject, it is more the driving force behind contemporary economic activities, the modalities and functioning of which it increasingly governs. Money transactions activities provide a good example of this dynamic. In the space of a decade, electronic and digital systems, through mobile payment, have fundamentally disrupted the national monetary ecosystem as well as the financial habits of Ivorians. The technologies that support these activities have transformed the sector from an archaic to a modern system. The migration from the old model to the new model is more an operational necessity than a functional market choice. The disruptive innovation brought by these systems has in turn enabled a new spatial structure of monetary activities, practically on the scale of the whole of Côte d'Ivoire, as well as the financial inclusion of many Ivorian populations, hitherto on the margins of the traditional financial circuit. Admittedly, in many rural communities, the permanence of so-called gray and white areas of electronic communication does not always facilitate transactions, to the point that the populations of these communities are still forced to travel to urban communities to conduct their operations. However, all in all, the advent of mobile payment has considerably changed the relationship of these populations with the financial circuit. Given its broad scope, the phenomenon of digital money transactions has imposed a new economic model (mobile payment), significantly reduced asymmetries in access to transactions and led to a new geography of financial activities in Côte d'Ivoire.

To understand the full implications of digital technology in society, two main lines of thought were given priority in this research: the analysis of the economic model of digital monetary transactions and the analysis of the spatial structures generated by these activities. It emerges that integrated into the composite fields of analysis (geographical, economic and sociological), digital tools appear to be powerful interpretative means to be studied. They are indeed real producers of geographical areas, factors of economic activity and social comfort. By accelerating the circulation of money, digital payment platforms, mainly mobile money, have created a unique and broad-based economic model in Africa. Without digital technologies and mobile payment, a large part of the Ivorian population would still be excluded from the modern financial circuit. Nevertheless, since the mobile payment revolution is still in its infancy, we may wonder, given the speed of change, how the money circulation model will evolve in the future. In any case, the reflection on the strategic challenges of digital technology for the development of territories in Côte d'Ivoire remains largely open, insofar as digital tools will continue to shape people's lifestyles, activities and achievements.

## 8.10. References

ARTCI (2018a). Statistiques du marché des télécommunications en Côte d'Ivoire [Online]. Available at: http://www.artci.ci/images/stories/pdf/rapport_activite/stat_3e_trim_2018.pdf document [Accessed 30 May 2019].

ARTCI (2018b). Chiffre d'affaires de la téléphonie mobile en Côte d'Ivoire. [Online]. Available at: http://www.artci.ci/index.php/Telephonie-mobile/chiffre-daffaires-service-mobile.html [Accessed 30 May 2019].

Banque mondiale (2018). Pour que demain ne meure jamais. La Côte d'Ivoire face au changement climatique. Rapport de la Banque mondiale sur la situation économique de la Côte d'Ivoire [Online]. Available at: http://documents.worldbank.org/curated/en/470341530853819903/pdf/127979-REVISED-WP-FRENCH-P167083-PUBLIC.pdf [Accessed 20 November 2018].

BCEAO (2018). Note d'information du quatrième trimestre 2018 [Online]. 56. Available at: https://www.bceao.int/sites/default/files/2019-02/Note%20d%27information%20-%20n%C2%B0%2056%2C%204e%20trimestre%202018.pdf [Accessed 30 May 2019].

Cerstin, S. *et al.* (2005). Étude sur le transfert d'argent des émigrés au Sénégal et les services de transfert en micro finance. Comment optimiser le rendement de transferts de fonds des émigrés vers leurs pays d'origine? International Labour Organization (ILO).

Christensen, C.M. (1997). *The Innovator's Dilemma*. Harvard Business School Press, Brighton.

Christensen, C.M., Raynor, M. (2003). *The Innovator's Solution*. Harvard Business School Press, Brighton.

Hussy, C., Mercier, C., Raffestin, C. (1985). Centralité et concentration [Online]. *Cahiers de géographie du Québec*. Available at: https://doi.org/10.7202/021691ar. 29(76), 9–28.

International Telecommunication Union (2018). *Measuring the information society*. Rapport 2017, 1.

Loukou, A.-F. (2013a). Les obstacles à la diffusion de l'Internet en Côte d'Ivoire, et ses implications. *NETCOM (Networks and Telecommunication Studies)*, 26(2012), (3-4), 307–328.

Loukou, A.-F. (2013b). Les systèmes de transfert électronique d'argent. Un exemple d'utilisation des TIC au service de l'aménagement du territoire en Côte d'Ivoire. In *Transformations sociales et ère numérique*. Montreal, Quebec.

Loukou, A.-F. (2016). Les enjeux sociaux et géographiques des banques numériques en Côte d'Ivoire. *Revue Échanges*, 2, 501–514.

Pumain, D. (2004). Les lieux centraux selon Christaller [Online]. Available at: http://www.hypergeo.eu/IMG/_article_PDF/article_13.pdf [Accessed 15 August 2018].

Rogez, O. (2018). La révolution du *mobile money* en Afrique: acte 1, la naissance [Online]. Available at: http://www.rfi.fr/emission/20181226-revolution-mobile-money-afrique-acte-1-naissance [Accessed 27 December 2018].

Sadin, É. (2013). *L'humanité augmentée. L'administration électronique du monde.* L'Échappée, Paris.

World Bank (2003). Global Development Finance. Striving for Stability in Development Finance. Document, World Bank, Washington DC.

PART 3

# Digital Technologies and Human Development

# 9

# Digital Technology, "Smart City" and Video Surveillance in Côte d'Ivoire

Nowadays, ICTs are essential, even structuring to life in society (Wolton 2000). Any sustainable development project includes ICTs and aims to address the multiple challenges faced by countries (poor and emerging). The objective of sustainable development at various levels of society is to establish digital infrastructures. In other words, digital technologies are extending their impact to broader areas of global transformation (Kiyindou 2019). Africa is not to be outdone by such a change as witnessed by the use of ICTs in an increasing number of development projects. These projects make it possible to develop, elsewhere as in Africa, more modern concepts aimed at improving the living conditions of populations in countries, regions, territorial and localities. This is particularly true in municipalities where "smart cities" projects, through the immersion of digital technologies in the daily lives of city dwellers, become development visions among decision-makers.

In order to mobilize resources against cybercrime and the growing insecurity such as the terrorism that has been rampant in the West African sub-region (Burkina, Mali, Nigeria, etc.) in recent years, the government of Côte d'Ivoire has taken several measures to provide itself with the means to ensure better surveillance of both cyberspace and the Ivorian physical territory. Since the end of the post-electoral crisis in 2011, these measures have consisted in the implementation of a large project to identify all mobile and Internet users. The project has also led to major State investments in new

Chapter written by Julien ATCHOUA and Jean-Jacques BOGUI.

technologies for territorial surveillance, communications and surveillance cameras in some Ivorian cities. It is with great pride that the government announced the installation of nearly 1,000 cameras in the city of Abidjan to ensure the security of this economic metropolis (Wassimagnon 2019).

For the Ivorian government, these achievements constitute a great source of pride with the official objective of fighting insecurity by repositioning Abidjan as a "smart city". However, an analysis of the reactions to acts of repression by the general population and Ivorian Internet users on social networks reveals a certain skepticism about the expressed will of the Ivorian state to ensure the well-being of its citizens. The introduction of information and communication technologies such as telecommunications networks into urban infrastructure is certainly a significant factor in achieving the development objectives prescribed by international organizations such as the United Nations. They also call for theoretical and pragmatic reflection on the socio-political and communication challenges presented by such facilities for the supposed main beneficiaries of these achievements, namely the citizens of the urban areas in question. From a methodological standpoint, the response to such a concern has led us to a qualitative analysis of a corpus of approximately 50 comments from Internet users and the use of about 10 press articles on the current state of digital surveillance facts reported by official online media and blogs. The information obtained makes it possible to understand the motivations linked to the need for the Ivorian authorities to include "digital intelligence" in their development projects. It also makes it possible to understand the challenges presented by digital installations for urban populations in Côte d'Ivoire.

## 9.1. Smart city in Côte d'Ivoire: prospects

As reported by the United Nations (UN) Department of Economic and Social Affairs since 2014, urban dwellers represent more than half of the world's population (Simard 2015). This concentration of populations in cities is as important as the problems they generate and embodies threats of all kinds to which governments, territories and citizens themselves are called upon to propose appropriate solutions, especially in terms of security (Bagnis and Barois 2017). Electronic means appear as "a tool that must be seized to think and develop the city of tomorrow. It represents tremendous opportunities to build or rebuild 'smarter' cities and territories and reinvent a new urban character" (Bagnis and Barois 2017). According to the UN as cited by Simard (2015), the smart *city*, which aims to respond to the societal and environmental challenges now facing the world, is the gateway to more secure urban cities, with increasingly growing urban populations (nearly 70% by 2050).

In light of such contemporary challenges where the use of information and communication technologies in various sectors of activity makes it possible to optimize the use of local infrastructures, this study examines the situation of Côte d'Ivoire as a West African country that intends to make ICT a part of its development while experiencing rapid growth in its urban populations.

In his opening speech in October 2018 at the seventh edition of the *Assises de la transformation digitale en Afrique* (ATDA) in Abidjan, the Ivorian Minister of the Digital Economy and Communications invited panelists from French-speaking countries (Benin, Burkina Faso, Senegal and France) to identify the problems of Africans in terms of "digital governance and infrastructure" (Akeko 2018) and showed that ICT is essential in Côte d'Ivoire's position as a country of the future. In 2017, the Ivorian government adopted a law enabling the State to engage in digital development by introducing ICTs into all sectors of activity. The objective was to ensure better living and working conditions for people. This marks the Ivorian State's desire to move towards development with the digital options and practices (Akeko 2018) that already seem to be underway. According to an article by Kobo, which reveals about the Ivorian economic capital Abidjan:

> "It is being digitized without the inhabitants really realizing it and there is no turning back. At the end of the post-election conflict in 2011, which claimed the lives of 3,000 people, the new Ivorian government implemented several ICT projects aimed at modernizing the city[of Abidjan]and reducing congestion due to the displacement of people who had fled war zones to Abidjan." (Kobo 2016, p. 52)

Abidjan would thus emerge as "an intelligent city", even if the concept remains unknown to city dwellers who live in an increasingly connected city, says Kobo. He believes that:

> "Information and communication technologies have changed the attitudes of a large part of the population of the city [of Abidjan], estimated at 4.7 million inhabitants, as well as the quality of the services provided. Long gone is the time when pensioners used to go to the bank with a raffia mat folded under their arms to sit on the floor during endless queues. Today, residents can send and receive money from their mobile phones, and even pay school fees. Those who know how to use the Internet and they can consult their children's test results before they even return from school." (Kobo 2016, p. 52)

These words reflect the digital development actions undertaken by the Ivorian State under the government's leadership following the controversial 2010 elections. The elections led to a politico-military crisis in 2011 after talks broke down between the warring parties. As a result, several sectors of activity were affected, including the more sensitive aspect of security through video surveillance in this movement towards the smart city. This has prompted the Ivorian government to install a thousand cameras in Abidjan at various strategic locations in the city. In an online article, a blogger who calls himself "Le Petit Kettinois" (September 2019) reveals through his writings "the seizure of power by surveillance cameras in Abidjan". These remarks reflect the effectiveness of the presence of digital control devices in the main streets of the Ivorian economic capital. Nevertheless, in a global context of post-electoral crisis, these facilities have provoked numerous controversies in public opinion about the real intentions of government and police authorities while stimulating scientific curiosity to understand them.

## 9.2. Video surveillance in Abidjan: state of controversy

The development of media technologies is the subject of contradictory perceptions (the Athens-Orwell dichotomy). The so-called Athenian vision perceives an association between media technologies, democracy, communication and social progress (Lévy 2002). This approach highlights the egalitarian and liberating aspects that these technologies can promote. From this perspective, the source of conflicts is generally perceived as being due to a lack of communication since the role of communication networks is to restore social cohesion. Gingras, in his definition of public space, makes the following observation:

> "The characteristics of electronic networks, such as access to information, interactivity, digitization and high-speed transmission, give rise to optimism that two of the most cherished ideals of democracy that directly concern the public sphere can be achieved: transparency and agora." (Gingras 2009, p. 216)

The transparency of the network is justified by access to information in a large quantity of documents (public documents, archives, research results, etc.). Agora is allowed because of the interactive nature of the system, which encourages exchanges between users (discussion groups, blogs, etc.) and freedom of expression because of the lack of control (Gingras 2009, p. 216).

However, while the Internet can be a tool for greater transparency in public action, advocates of the Orwellian vision highlight the potential for monitoring and surveillance of ICT. Today, citizens' personal information is easier to acquire because of the Internet (knowledge of identity, browsing habits, consumption, etc.). ICTs can also be used to detect behaviors that may be considered dangerous for society or for tracing and recording citizens based on information about their tastes, habits, political or ideological backgrounds, etc. Thus, far from promoting transparency, digital technologies are more likely to be used as instruments to facilitate espionage and profiling of populations. The information society would then be a society of control and supervision.

Surveillance is theoretically a major asset in the fight against insecurity and has intensified in recent years in our contemporary societies with the development of ICTs. Today, it has such a hold on the daily lives of citizens that it is almost impossible to escape it. Lyon (2001, p. 1) expresses this situation well in the following terms: "Everyday life is subject to monitoring, checking, scrutinizing. It is hard to find a place, or an activity, that is shielded or secure from some purposeful tracking [...]."

Bennett *et al.* (2014, p. 7) define this notion of monitoring as "any systematic special interest in personal information to influence, manage, authorize or control the persons to whom the information collected relates". Some authors also speak of population profiling, which, according to Mattelart and Vitalis (2014, p. 5), is defined as:

> "[...] a form of indirect control of individuals through the use of information collected from them. From the workers' logbook to police registers and manual files, up to the emergence of information technology and the Internet, this form of control never ceased to improve and expand. From law enforcement to banking, from intelligence to marketing, from educational institutions to psycho-medical services, few sectors of activity are exempt from this."

In the case of Abidjan where cameras emerged for the safety of citizens according to the official discourse justifying their installation, an online cartoonist writes:

> "For some time now, strange 'eyes' have been hanging from heights all over the city of Abidjan. Intrusive, they probe, scan, deter and warn in a way similar to the voyeuristic trait of Karaba the witch in the animated film 'Kirikou'. As you may have guessed, these are the surveillance cameras that swarm the streets and other hot spots of Abidjan." (Le Petit Kettinois 2019)

Nearly 1,024 surveillance cameras are installed in the city of Abidjan, which represents 91.6% more installations than in 2018. For example, according to the Ministry of the Interior[1], they identified 821 traffic accidents on public roads.

These installations, which appear to be useful for security management or control by road surveillance centers (traffic jams, physical assaults, digital criminal tracing, etc.), give rise to differing opinions among Internet users on the suspected or proven motivations of the political and police authorities that control them.

## 9.3. Surveillance cameras between hopes and fears

It appears from the comments of Internet users that video surveillance in Abidjan constitutes a "miracle and incomparable solution for the security of populations". Security is understood as "a state in which hazards and conditions that can cause physical, psychological or material damage are

1 According to a July 2019 info alert from the Ministry of the Interior.

controlled in a manner that preserves the health and well-being of individuals and the community" (Kiyindou 2019, p. 146). We note that on the Internet and in the various forms of expression used by the views on securing Abidjan through video surveillance, some Internet users express feelings of satisfaction because they consider that security will be much better than before digital surveillance installations. It would come at the right time according to the following words of an Internet user, identified under the pseudonym "Mack Javel":

> "I think, however, that these installations will help to significantly reduce insecurity in our cities. It will only be necessary to identify hot spots, major junctions, strategic points on the highways and all over the country. If these cameras are used properly, the 'microbe' phenomenon, robbery and road blockers will be distant memories."

In addition, one category of Internet users is more skeptical about the execution of this large project of a thousand cameras to be installed by the Ivorian government. This leads to a contradiction in views on this public security policy. In his commentary, the Internet user DB invites all commentators to see in these Ivorian government works a social welfare service and to make efforts to avoid treating these achievements as espionage instruments. We believe that this is not realistic. He translates his words in these terms: "We are talking about securing the country, you are talking about spying!!!!! Are you alright? [...]"

While this Internet user is relatively optimistic about the role of cameras in Abidjan, he is still influenced by the point of view of an Internet user who is pessimistic about the rationale for these new-generation devices. The latter raises the issue of spying on citizens by the competent authorities (governmental and police) in order to restrict their freedom and consequently to silence them on sensitive subjects of socio-political life. This Internet user questions the acceptability of monitoring citizens in order to deprive them of part of their freedom, which is one of the foundations of a deliberative process of a participatory democracy. The fears of Internet users who argue in the same sense as the previous one are related to the use of the data that will be collected by these cameras and their accessibility to the populations. Data that, moreover, are information about their private lives. They worry about a probable interference with collective and individual freedoms, which cameras would make possible. Transparency in the management of this information is another concern in the comments collected.

In any case, the deployment of video surveillance as a security device seems to be perceived favorably or not depending on whether an individual is a supporter of the current government or the opposition. On the Internet, comments and opinions are rife and express two points of view. On the one hand, they consist of highlighting the general interest sought by governments, i.e. public security. And on the other hand, they raise the question of governance and the long-term maintenance of this system, which would require major public investments since it is copied from models. Video surveillance security is therefore a growing trend. Industrialized countries are used as an example to justify African countries' choices to use this technique to ensure the safety of their citizens:

> "Do you know what precautions they took? You do too much 'copy/paste' all the time as if the variables were the same everywhere. This is also the underdevelopment and utopia of emergence."

Through such writings, this Internet user questions the Ivorian government on the precautions taken to ensure the model can be reproduced in the installations. For them, it is not suitable to copy a development model. They consider that all development is unique and takes place according to the realities of each country. Wanting to appear as an emerging country by applying a Western model therefore does not bode well for a reassuring future.

The Internet user is also concerned that the Ivorian State has entrusted the Abidjanese digital installations to a foreign company, in particular a Chinese company, whose espionage possibilities they fear. At the same time, they discredit the company in charge of implementing the project. Through their reflections, this Internet user wants to be a lesson provider based on what they think of their country.

It should also be noted that the implementation of public policies for the construction of a "smart city" is not without disappointment. It is entirely legitimate for people to monitor the use of public funds. If acts of embezzlement of public funds are identified, people are entitled to hold governments to account. This contributes to improving the governance of public projects. Even if the advent of a healthy project is positively appreciated, the fact remains that its implementation cannot be achieved without incriminating acts on the part of the sponsors. The companies responsible for implementing these projects are also not subject to the same incriminating acts. The "yes, but" is the expression which characterizes this mistrust on the part of the population.

## 9.4. Conclusion

The development of media technologies is the subject of contradicting perceptions (Athens-Orwell dichotomy). The so-called Athenian vision perceives an association between media technologies, democracy, communication and social progress (Lévy 2002). This approach highlights the egalitarian and liberating aspects that these technologies can promote. From this perspective, the source of conflicts is generally perceived as being due to a lack of communication, as we saw in Gingras (2009) with the role of communication networks being to restore social cohesion. At this level, action plans consist of governments and other competent authorities taking an interest in digital technologies in urban installations or, if necessary, giving them greater importance in the solutions to be proposed to deal with the environmental, economic and socio-political contingencies of urban cities. Initiatives such as these, recalls Cazabat (2018, p. 22), "have definite potential, but they are still too marginal to improve development indicators at the global or even national level". Urban technological innovation as an incubator for a "smart city" then constitutes a footprint on the overall modernization of municipalities in Africa in terms of actual achievements or development projects.

In short, the development of ICTs makes it possible to improve human activities in almost all fields. More and more governments are giving these tools a prominent place to maximize their intervention for the benefit of populations. The security sector is benefiting from this technological advancement. For security reasons, there has been a significant deployment of various surveillance devices on the streets of large urban areas in both northern and southern countries. Côte d'Ivoire is not on the fringes of such a switchover to digital in the overall project to build *smart cities*. The main systems of the road network in Abidjan, the Ivorian economic capital, have been equipped with various kinds of digital equipment since 2017. We can note the presence of pylons, cameras and many other latest generation devices.

From now on, the actions of the Abidjanese people are watched by video surveillance devices on every street corner. This has provoked reactions from the population. While some believe that the government is making a good choice by investing in digital surveillance, others are less enthusiastic about the real intentions of the leaders and the opportunity for such major work in a country that is poor (46.3% poverty rate – World Bank, 2019), highly indebted (PPTE) and politically sensitive. These verbal trends were translated into online comments on the latest developments in video surveillance in Côte d'Ivoire and into information from interviews with competent authorities, allowing the analysis to be more rigorous in its interpretation of points of view on the issue. Opinions on technological achievements on Abidjan's public roads range from issues related to public and individual freedoms in a democratically harmful environment, to security issues against large-scale street banditry and to the management or governance of achievements.

While some consider these urban installations as political instruments of population control, or even possible abuse of power by public authorities who would use them to settle scores with their opponents, others perceive these achievements as an improvement in the populations' safe living environment. Referring to the socio-political electoral crisis that the country has experienced, the second speeches collected maintain the idea that video surveillance makes it possible to end the attempts to destabilize Côte d'Ivoire by vengeful minds, i.e. those of the victims who would be tempted to take justice into their own hands through means not stated. Internet users are also concerned about the durability of these installations, which are perceived as models imported from Western countries with more means of protection than Côte d'Ivoire. Their country thus seems to them to be forcing the achievement of the development objectives of an "emerging country by 2020", according to the political statements of the new Ivorian power resulting from the 2010 presidential elections.

It is certainly common to observe a counter-power in the implementation of public policies, increasingly expressed online in the form of contradictory discourse. Video surveillance as one of the obvious elements in the construction of a "smart city" to achieve the objectives of an emerging Côte d'Ivoire is far from reassuring a category of inhabitants who are expressing themselves on the Internet as a means of security. Is it a problem of adaptation to tools that are now part of their daily lives or something else? This exploratory work aims to open up such perspectives, the answers to which would make it possible to gain an in-depth understanding of the discourses of urban populations regarding these installations.

## 9.5. References


Akeko, A. (2018). ATDA Abidjan 2018: villes intelligentes, comprendre l'approche africaine [Online]. *Cio mag*. Available at: https://cio-mag.com/cote-divoire-le-ministre-isaac-de-ouvre-les-atda-2018/ [Accessed 5 December 2018].

Alerte Info (2019). Abidjan, la ville aux 1 000 caméras de surveillance [Online]. Available at: http://www.atoo.ci/2019/07/22/abidjan-la-ville-aux-1-000-cameras-de-surveillance/ [Accessed 8 September 2019].

Bagnis, M. and Barois, B. (2017). 10 minutes sur la *smart city*, une ville intelligente. Des villes et des territoires plus inclusifs, durables et pratiques. Document, Agence d'urbanisme de l'aire toulonnaise et du Var, Ollioules.

Banque mondiale (2019). Côte d'Ivoire Présentation-Banque mondiale [Document]. Available at: banquemondiale.org [Accessed 15 September 2019].

Bennett, C., Haggerty, K., Lyon, D. and Steeves, V. (2014). *Vivre à nu, la surveillance du Canada*. Athabasca University Press, Edmonton.

Bogui Maomra, J.-J. and Atchoua, J. (2016). La régulation des usages des TIC en Côte d'Ivoire: entre identification et craintes de profilage des populations. *tic&société/Terminal*, 10(1), 118.

Cazabat, C. (2018). L'impact des technologies de l'information et de la communication sur le développement humain. In *Objets connectés et développement intelligent*, Kiyindou, A., Akam, N. and Damone, E. (eds). L'Harmattan, Paris, 13–26.

Gingras, A.-M. (2009). *Médias et démocratie: le grand malentendu*. Presses de l'Université du Québec, Quebec.

Kiyindou, A. (2019). *Intelligence artificielle. Pratiques et enjeux pour le développement*. L'Harmattan, Paris.

Kobo, K. (2016). Abidjan, l'émergence d'une ville intelligente [Online]. *Villes d'Afrique, Le nouveau magazine de CGLU-A*, 5, 52–55. Available at: https://www.google.com/url?sa=t&rct=j&q=&esrc=s&source=web&cd=1&ved=2ahUKEwjR06izyL_lAhWJi1wKHVbFBFsQFjAAegQIAhAC&url=https%3A%2F%2Fwww.uclga.org%2Fwp-content%2Fuploads%2F2018%2F02%2FAfricaines-Villes-05.pdf&usg=AOvVaw1U8TjjY8p6wf__6VvqCozr [Accessed 8 September 2019].

Le Petit Kettinois (2017). La prise de pouvoir des caméras de surveillance à Abidjan [Online]. Available at: http://lepetitkettinois.mondoblog.org/2017/09/29/abidjan-ville-camera/ [Accessed 8 September 2019].

Lévy, P. (2002). *Cyberdémocratie: essai de philosophie politique*. Odile Jacob, Paris.

Lyon, D. (2001). *Surveillance Society: Monitoring Everyday Life*. Open University Press, Buckingham/Philadelphie.

Mattelart, A. and Vitalis, A. (2014). *Le profilage des populations, du livret ouvrier au cybercontrôle*. La Découverte, Paris.

Simard, J. (2015). La ville intelligente comme vecteur pour le développement durable: le cas de la ville de Montréal. Master's thesis, University of Sherbrooke, Sherbrooke.

# 10

# The Integration of ICT in Higher Education in Senegal

From the beginning, information technology has played a key role in almost all areas of human life. Similarly to the Global North, developing countries have entered the digital age. Thus, we are witnessing an exponential growth in the use of ICT.

ICT can be used in many areas of daily life. Whether accessing knowledge, leisure, banking, etc., they provide a multitude of solutions to the problems we face every day.

Thus, to achieve the MDGs and universal schooling, Senegal has opted for the integration of ICTs into education. This integration has led to the implementation of e-learning.

The objective of our work is to take stock of the state of e-learning in Senegal. For this purpose, first, we look at the current situation of ICT in Senegal. Second, we provide some background to the integration of ICTs into education. Third, we outline the context and challenges of e-learning in Senegal. Fourth, we describe e-learning initiatives in public and private higher education institutions. Finally, we list the advantages, disadvantages and problems encountered by e-learning in Senegal.

---

Chapter written by Ousmane SALL, Mouhamadou S. DIALLO, Insa BADJI, Moustapha CISSÉ, Abdoulaye SY and Marie NDIAYE.

## 10.1. State of ICT in Senegal

Since its inception, ICTs have played a leading role in almost every aspect of life around the world.

Like sub-Saharan countries, Senegal attaches great importance to the use of ICTs (Fournier Fall 2006). This has taken the form of a new telecommunications code and created a telecommunications and post-regulatory agency (*Agence de régulation des télécommunications et des postes*, ARTP), a state information technology department (*Direction informatique de l'État*, DIE) and a ministry in charge of post and telecommunications.

Thus, we are witnessing a growing desire to use ICT for administration purposes. To this end, several measures have been taken, including (Sy 2016):

– the definition of a national ICT development strategy;

– the adoption of a new telecommunications code;

– the creation of the telecommunications regulatory agency (*Agence de régulation des télécommunications*, ART);

– the creation of the state computing agency (*Agence de l'informatique de l'État*, ADIE);

– the creation of a ministry in charge of promoting ICT;

– the full liberalization of the telecommunications sector.

Given the importance of ICTs and the contribution they have made for several decades now, Senegal has opted for the integration of ICTs into education.

## 10.2. Integration of ICT into education

For more than 20 years now, the integration of ICT into education has become a reality in Senegal, which has been motivated by several factors. The most convincing ones are the improvement in the quality of education and the ease of access to information. However, we can add the desire to introduce students to an early use.

It all began in 1990 with the birth of the School Computer Generation (GIS). This was followed by the massive introduction of computers in schools, where each school had at least one computer room and introduced compulsory computer literacy courses.

In the spirit of continuity, the Ten-Year Education and Training Plan was launched in the 2000s. Its aim was to increase access to education.

Under the IT Master Plan (IMP), all schools were connected to the Internet, which reflects the desire to use ICTs for educational purposes. Since then, ICTs have been a lever for improving teaching and learning practices.

Established in 2003, the National Commission for the Integration of Information and Communication Technologies in Education (*Commission nationale pour l'intégration des technologies de l'information et de la communication à l'éducation*, COMNITIC), in close collaboration with the Information Technology Unit of the Ministry of Education (*Cellule informatique du ministère de l'1Education,* CIME), is responsible for leading the reflection on the pedagogical and didactic aspects of learning related to the use of IT tools in schools. In this way, it lays the foundations for multimedia teaching, e-learning and research through multimedia, which represent enormous assets for all those who seek to know more.

The SENECLIC project, which was launched in 2005, aims to reduce the digital divide and social exclusion in schools, which are still present in Senegal, by equipping elementary schools with multimedia rooms. This project connects more than 8,000 elementary schools to ICTs and has four specific objectives, namely (Oasis Media Group 2010):

– to establish a multi-stakeholder partnership for the proper collection of new and second-hand quality IT equipment;

– to popularize and develop digital education in elementary schools by installing multimedia rooms and training teachers to better teach students with digital tools;

– to strengthen the system for the reconditioning of computer equipment for e-waste management;

– to operationalize e-cases and the digital solidarity network.

The SANKORE program aims to equip schools and train teachers in the use of ICT.

As with elementary and middle schools, programs have been set up to facilitate the use of ICTs in higher education. At this level, the responsible minister, the World Bank and universities have committed themselves to developing a national network to interconnect universities. We are also witnessing an increased awareness regarding the usefulness of e-learning. From this point, the foundations of e-learning in Senegal began to be laid. In the following, we will present the context, challenges, advantages, disadvantages and problems encountered by e-learning in Senegal.

## 10.3. E-learning in Senegal

The use of ICTs has been very successful in Senegal. Their integration into the higher education system has become essential, prompting many to reflect on how ICTs could improve the quality of higher education and solve the problems that Senegalese universities face. Several solutions have been proposed, but the one that attracted our attention the most is the introduction of distance learning (e-learning).

In this section, we provide some background information and present the challenges regarding e-learning in Senegal.

### 10.3.1. *Background information*

Beyond the use of ICTs, the increasingly targeted objective is the media coverage of education, which ranges from the simple use of a computer or software in a course to the implementation of an e-learning platform (Lat and Sidir 2015). In our work, we focus on the Senegalese context.

Senegal is one of the first countries in sub-Saharan Africa to have taken initiatives to promote the development and use of ICT in education. It is in this sense that the current Minister of Higher Education, Research and Innovation from 2012 to 2019, Mary Teuw Niane, stated that:

> "We must show our commitment by taking initiatives that avoid marginalizing us on the path to development. We must take the turn that leads to development and equip ourselves with means

that are in line with the strengthening of pedagogy in all its forms."

Senegal has a small number of universities. Of the fourteen regions of this country, only five have a university with a low admission capacity. With the increasing number of students succeeding in the baccalaureate, it is becoming increasingly difficult for the State to guide new baccalaureate students. Moreover, these universities clearly lack pedagogical facilities such as lecture halls, classrooms and laboratories (Sy 2016). In addition, people in the most remote parts of the country who are admitted to these universities face many difficulties, often due to financial and geographical barriers. Moreover, in the context of international competition and the expansion of local education, the modernization of the education system has become essential.

All these elements mean that the State of Senegal, through its Development Plan for Higher Education and Research (PDESR), has for several years now taken the initiative to develop e-learning. The establishment of the *Université virtuelle du Sénégal* (UVS) and its open digital spaces (ODS) is a decisive step towards achieving these objectives (Gueye 2015).

### 10.3.2. *Issues*

To solve the problems facing higher education and the modernization of its education system, the State of Senegal has opted for the development of e-learning (Fournier Fall 2006). It is in this context that we have witnessed the creation of the UVS, which was perceived as a message from the State of Senegal in its desire to publicize higher education. The stakes are high, which consist of:

– *reducing university overcrowding:* by allowing students and professionals to follow e-learning courses;

– *removing geographical barriers:* by allowing everyone to pursue the training of their choice, regardless of where they are located;

– *instilling in students a research philosophy by promoting self-learning:* by having tutors to always guide them;

– *reducing spending:* by reducing the number of infrastructures to be built and by reducing spending on the management of public and private universities.

Despite the many challenges they face, e-learning has proved its worth and is increasingly gaining the trust of pupils, students and professionals.

In the following section, we will introduce the e-learning initiatives in higher education institutions in Senegal.

## 10.4. E-learning initiatives in higher education institutions in Senegal

In this section, we first present e-learning initiatives in public institutions and then outline what is being done in private institutions. This section is mainly based on the work of Abdoulaye Sy.

### 10.4.1. *E-learning initiatives in public institutions*

#### 10.4.1.1. *Case of the Université Cheikh Anta Diop (UCAD) in Dakar*

Several UCAD faculties offer e-learning courses. We will focus on two of these larger faculties, namely the Faculty of Science and Technology, Education and Training (FASTEF) and the Faculty of Legal and Political Sciences (FSJP).

The FASTEF is one of the first university institutions in the country to develop e-learning since 1980, when it began to offer e-learning for school principals. Since then, it has provided training in many areas such as administration, the environment and communication. Since 2009, the quality of the e-learning courses it offers has earned it the privilege of supporting the Ministry of National Education in the distance training of its contractual and part-time teachers in the middle and high school sectors, as well as of administrators. In three years, it has trained nearly 9,000 middle and high school teachers.

Each year, the FSJP faces a high entry demand by new graduates and professionals who wish to strengthen their knowledge of law. To meet these requests, it set up an e-learning program, which was bimodal, during the

2010–2011 academic year. The courses are available on the e-learning platform, but tutorials and evaluations must be done in person.

#### 10.4.1.2. *Initiatives of the Université virtuelle africaine (AVU)*

The AVU was established in 1990, which is one of the precursors of e-learning in Senegal. With the use of ICT, it has been able to set up an e-learning system, mainly offering scientific and technical training.

Its strategic objectives are to facilitate access to open, distance and e-learning resources that are of high quality and adapted to Africa. It has trained more than 40,000 students across Africa.

#### 10.4.1.3. *Case of the Université Gaston Berger (UGB) of Saint-Louis*

The UGB embarked on e-learning in 2003. It all started with a co-degree with Laval University in Canada for some courses in computer science, law, mathematics, etc.

To facilitate access to UGB e-learning, in 2010, we witnessed the creation of the Open and E-learning Institute (IFOAD).

The IFOAD is taking advantage of its cooperation with major international organizations such as USAID to equip itself and set up all the logistics to carry out their lessons online.

#### 10.4.1.4. *Initiatives of the Université virtuelle du Sénégal (UVS)*

The UVS is by far the most successful example in terms of e-learning in Senegal. It primarily offers e-learning on a dedicated platform.

Its system is based on a network of more than 100 teachers and more than 200 tutors (with all disciplines combined), as well as on open digital spaces (ODS), which enables students to fully understand that they are part of a community.

To support students, the State has set up the "Sama PC" program in order to equip all UVS students with a laptop and Internet connection keys (Dieng 2016).

## 10.5. E-learning initiatives in private institutions

Most of Senegal's private universities offer e-learning. As an example of a study, we considered the *Institut africain de management* (IAM) and the *Institut supérieur de formation en management, informatique et télécommunications* (MIT).

### 10.5.1. *IAM initiatives*

The IAM group primarily offers online training for professionals and students. To enable students to better understand the courses, the teaching team offers learning modules, which are divided into chapters and reinforced by articles or podcasts.

Each chapter ends with quizzes, MCQs or case studies as a knowledge test before starting another chapter. Support is also offered to the online community for a better understanding of their training.

### 10.5.2. *MIT Initiatives*

The MIT provides its students with a functional platform and relevant course materials designed by competent teachers and adapted to e-learning. To make it easier for students to understand the course, the MIT provides tutors.

Since the training provided is practical and oriented towards professional know-how, a multitude of activities are offered to students, including workshops, practical work and projects.

After having reviewed the current state of e-learning in Senegal, we will now turn our attention to its advantages, disadvantages and the problems it faces.

## 10.6. Advantages, disadvantages and problems encountered by e-learning in Senegal

As with most of the solutions proposed to solve the problems encountered, e-learning is no exception. Despite its enormous advantages, e-learning has disadvantages and faces several problems in Senegal.

In the following sections, we first present the advantages of e-learning in Senegal and then outline the disadvantages and problems faced by e-learning.

### 10.6.1. *Advantages of e-learning in Senegal*

E-learning offers the opportunity for pupils, students and professionals to pursue training courses, wherever they may be. This trend has spread all over the world to mark the many constraints encountered by face-to-face training. Thus, Senegal, in its desire to modernize its higher education and achieve universal schooling for all, has been increasingly taking advantage of the benefits offered by e-learning in recent years.

The most successful example in the field of e-learning is the *Université virtuelle du Sénégal* (UVS). Thus, to highlight the advantages, disadvantages and problems encountered by e-learning, we will focus mainly on the UVS.

The numerous advantages of e-learning are presented below.

In Senegal, we have seen a plethora of new baccalaureate graduates increase over the past few years. Faced with this, both public and private universities are finding themselves under an ever-increasing weight of student numbers. In addition, it is becoming increasingly difficult for the State to guide all new graduates. Thus, the State has established a policy that has led to the establishment of the UVS to guide thousands of students in it, thereby making it possible to considerably reduce the number of students in Senegalese universities that offer face-to-face training.

These universities consider two main components that are essential for their proper functioning: the social component and the pedagogical component. The social component requires the establishment and management of a social campus, which is very expensive for the State. This component also deals with food, accommodation, transportation and other means to improve the living conditions of students. The pedagogical aspect of higher education in Senegal involves building of classrooms, lecture halls, laboratories and administration offices, where the cost of classroom teaching is excessively high for the State. Thus, e-learning provides solutions to these problems by reducing costs.

E-learning also reduces disruptions in the higher education community who are located in different places. Another advantage of e-learning is that it promotes autonomy by allowing students to go beyond their limits and instilling in them the philosophy of research. Thus, e-learning allows students to be self-taught.

E-learning is also an effective and unobtrusive way to carry out other activities and attend classes at the same time. At the end of the training, there is no difference between diplomas. This learning technique benefits not only professionals, but also those with families. For professionals, e-learning makes it possible to fulfill their professional duties and to have the opportunity to follow the training of their choice. For those with families, e-learning allows them to attend classes while remaining close to their families.

Most people have time or distance problems, i.e. the places where they have to study their courses are far away, making it difficult to follow the training of their choice. E-learning provides solutions to these problems by allowing them to pursue the course of their choice, wherever they are, as well as to learn at their own pace. Thus, e-learning has the advantage of breaking down geographical barriers.

A final advantage is the ability to internationalize e-learning, which allows foreign students to take the courses remotely, without spending as much money to attend the courses. It can also further have sub-region universities so that students from various countries can obtain the same skills, thereby no longer requiring overseas qualifications equivalence studies.

Despite these many advantages, e-learning in Senegal has several disadvantages compared to face-to-face learning and faces enormous difficulties.

### 10.6.2. *Disadvantages and problems encountered by e-learning in Senegal*

Even though the implementation of e-learning is less expensive than face-to-face training, it faces difficulties and has significant disadvantages.

The main disadvantage of taking an e-learning course is isolation, making students feel lonely and not knowing where to turn if there is a problem. As a result, students have to make huge efforts to absorb the content of the courses, and they are driven by a sense of loneliness which is often accompanied by deep discouragement.

Moreover, students become autonomous as there is no one to supervise their work. Thus, if this autonomy is poorly managed due to the lack of supervision, students can easily find themselves overwhelmed by courses and homework, and may even miss discussion sessions with their tutors. In order to succeed in e-learning, students need motivation, a strong mind and a great deal of commitment, because there is little or no contact with tutors. Therefore, it is important to judge the ability of students to learn independently before committing to e-learning.

In addition, as e-learning is based on the use of a platform, it is necessary to have a good understanding of the IT tool, without forgetting the possible difficulties of use. However, in the African context, the digital divide is still relevant and Senegal is no exception. Most learners take an interest in ICT, or become aware of its usefulness when they reach the level of higher education.

One of the major problems when taking an e-learning course in Senegal is Internet connectivity. A good Internet connection is an essential link in the development of e-learning. Thus, in the Senegalese context, many areas either have very low Internet connection speeds or do not have access to it. Students in these areas face enormous difficulties in properly carrying out their training, and, in most cases, end up throwing in the towel.

For essentially practical training, there is a risk of expanding on the theoretical aspect without enough practical experience. For this reason, in Senegal, face-to-face training is given more credibility than distance training. This is true despite the fact that for all private higher education institutions, the cost of e-learning is higher than face-to-face training.

Despite the many disadvantages and problems mentioned above, e-learning is increasingly gaining the trust of people in Senegal.

## 10.7. Conclusion

In Senegal, e-learning – which has been made possible by the development of media coverage – has emerged as a solution to many problems facing higher education, providing learners with the increased and varied possibilities of a remote education. Media coverage ranges from the simple use of a computer or software in a course to the implementation of an e-learning platform. This media coverage process has been accelerated by the State's desire to strengthen pedagogy in all its forms.

We cannot talk about media coverage without referring to the development of ICTs. With this in mind, we presented the current situation of ICTs in Senegal. ICTs have played a major role in the development of most sectors of human activity, modernizing teaching and learning methods. From the beginning, ICTs have been imposed as key tools to increase the quality of education and facilitate access to knowledge.

Senegal, like the sub-Saharan countries, in its move towards development, has found ways, through programs and cooperation, to integrate ICT into education. The willingness of Senegal to use ICT dates back to the 1990s.

Thus, to achieve universal schooling for all and solve problems facing its higher education, the Senegalese government has opted for e-learning.

In this work, we have presented the context and issues, as well as showed e-learning initiatives in public and private higher education institutions. This allowed us to identify the advantages, disadvantages and problems encountered by e-learning in Senegal.

Despite the many problems faced by e-learning in Senegal, it continues to gain the trust of pupils, students and professionals, as well as to achieve its objectives.

## 10.8. References

Camara El Hadji, H. (2011). L'usage didactique des technologies de l'information et de la communication (TIC) par les professeurs de l'enseignement moyen et secondaire au Sénégal et au Mali. *Liens*, 14, November.

Dieng, S. (2016). Analyse de la pratique du tutorat dans l'enseignement à distance: cas de l'université virtuelle du Sénégal. Master's thesis, University of Thiès, Thiès.

Diop, M.L. (2015). *E-learning* dans l'enseignement supérieur au Sénégal, entre succès et tensions: cas de deux établissements de l'université de Dakar [Online]. Available at: http://www.adjectif.net/spip/spip.php?article352.

El-Bouazzaoui, M. (2012). Les enjeux de l'intégration des TIC dans l'apprentissage de la langue étrangère. *Revue électronique internationale de sciences du langage*, June.

Fournier Fall, A. (2006). Enseignement à distance supporté par les NTIC au Sénégal. PhD thesis, University of Fribourg, Fribourg.

Guéyé, P. (2015). L'enseignement à distance et les MOOC au Sénégal: état des lieux et problématique. Colloque. In *Nouveaux défis de l'assurance qualité ; des ESG 2.0 aux MOOC*. Nancy.

Lat, M., Sidir, M. (2015). L'intégration des TIC dans l'enseignement supérieur au Sénégal: émergence de l'hypermédiatisation. In *H2PTM'15: Le numérique à l'ère de l'Internet des objets, de l'hypertexte à l'hyper-objet*, Saleh, I. (ed.). ISTE Editions, London.

Lishou, C. (2004). Usages de la formation à distance dans l'enseignement supérieur au Sénégal. Bilan et stratégies de mise en œuvre, Université Cheikh Anta Diop, Dakar, October.

Ndiaye, N. (2011). Les technologies de l'information et de la communication et l'enseignement à distance dans un environnement de massification des effectifs d'étudiants: le cas de l'Université Cheikh Anta Diop de Dakar (UCAD). PhD thesis, Université de Bordeaux 2, Bordeaux.

Oasis Media Group (2010). Réduction de la fracture numérique par une meilleure intégration des TIC à l'école élémentaire à travers le recyclage utilitaire et durable des équipements informatiques. Final report, Oasis Media Group, 17 October.

Sehi, B., Mian, A. (2019). Panorama des usages pédagogiques des médias sociaux dans l'enseignement supérieur en Afrique. *Africampus*, 9 May.

Sy, A. (2016). Intégration des technologies éducatives au Sénégal: les enjeux et les politiques (infrastructures, formation et les contenus). PhD thesis, University of Thiès, Thiès.

## 10.9. Webography

https://www.ucad.sn/index.php?option=com_content&view=article&id=1314&Itemid=396 [Accessed 15 June 2018].

https://www.usherbrooke.ca/biblio/accueil/biblio-manchettes/biblio-manchettes-details/article/34751 [Accessed 15 June 2018].

https://www.service-public.fr/particuliers/vosdroits/F541.a_formations_a_distance_avantages_et_inconvenients [Accessed 15 June 2018].

https://www.journaluniversitaire.com/formation-a-distance/ [Accessed 15 June 2018].

https://letarif.com/cours-particuliers/les-cours-distance [Accessed 15 June 2018].

https://www.rewmi.com/introduction-doutils-numeriques-lecole-enjeux-defis-au-senegal- bira-sall.html [Accessed 15 June 2018].

# 11

# Connected Objects in Africa: The Case of Telemedicine with Genesis Telecare in Cameroon

Telemedicine is defined by the World Health Organization (WHO) as the activity that "makes it possible to provide health services, where distance and isolation are a critical factor, by professionals (especially doctors) using ICTs for diagnosis, treatment and prevention, research and continuing education"[1].

It has been officially practiced since 1920, the year of the first license for a medical service radio aboard boats published in New York, USA[2].

It is taking its first steps in Africa, on a continent where many people are struggling with their health, due to living mainly in isolated areas. In addition to these human resource difficulties, there is a lack of infrastructure and equipment and limited purchasing power for a large proportion of the population. Such limitations (lack of quality human resources, lack of infrastructure, poverty) put the health of thousands of Africans at risk.

Under these conditions, telemedicine (healthcare and services, social, preventive or curative services, delivered remotely through ICTs) has the

Chapter written by Gilbert TOPPÉ.

1 Preamble adopted by the International Conference on Health, New York, 19–22 June 1946 and signed on 22 July 1946 by representatives of 61 States.

2 See (Amenta 2005).

potential to reduce inequality in the quality of healthcare services between urban and rural areas. It also supports isolated health workers in rural areas by providing them access to a solid medical knowledge base. Genesis Telecare is experimenting with this medicine in Cameroon.

This chapter first presents some general information on the functioning of telemedicine and then focuses on the specificities of the African case in Cameroon. Finally, it proposes a table of technological innovations that Africa needs to develop for this type of medicine.

## 11.1. General information on telemedicine

"Telemedicine is a form of remote medical practice using information and communication technologies[3]." It includes medical practices permitted or facilitated by telecommunications. It is a medical practice that allows for remote health services and the exchange of related medical information. It connects one or more health professionals with each other or with a patient, and among them includes a medical professional and, where appropriate, other professionals (psychologists, sociologists, etc.) providing care to the patient.

It makes it possible to establish a diagnosis and ensure follow-ups for preventive or post-treatment purposes for patients at risk. It also supports requesting expert advice, preparing treatment decisions, prescribing products, performing services or procedures and monitoring the condition of patients remotely.

Studies conducted by Teladoc[4] have shown that its service provides easy access to care at a lower cost and greater customer satisfaction. More importantly, access to telemedicine reduces the number of unnecessary visits to emergency rooms, where resources are scarce and costs can quickly accumulate[5].

This type of medicine represents a subset of telehealth, focusing on the use of information and communication technologies (ICTs) as part of

---

3 Organisation de la téléradiologie – Guide pour le bon usage professionnel et déontologique de la téléradiologie – Conseil professionnel de la radiologie.

4 Teladoc is a telemedicine provider in the United States (since 2002).

5 Available at: https://www.theepochtimes.com.

a program of care. It is manifested in many innovative applications and technologies: high-definition cameras, digital stethoscopes, biometric sensors, etc. In doing so, it allows the remote medical follow-up of a patient through these applications (Bonjao 2011).

At an organizational level, this practice allows hospitals to remedy the problem of medical desertification and thus offer patients the possibility of regular access to appropriate care without necessarily having to travel physically or waiting a long time to see the first available practitioner. Indeed, cost control and time-saving maximization are some of the direct benefits of this type of medicine[6].

In the case of medical isolation, for example, a patient or family member may transmit important information to an attending physician via the Internet or telephone. This allows physicians to confirm the declared medical case and provide emergency treatment advice or other guidelines for the care (medical, psychological, etc.) of the patient.

The practice of telemedicine gives doctors the opportunity to make reliable diagnoses quickly by transmitting more precise information via a high-definition camera, a digital camera, a smartphone, a tablet connected to the Internet or other indicator sensors that can help establish an expected diagnosis.

Although telemedicine can be an advantageous practice, there are limitations. Therefore, it cannot be self-sufficient and must be complemented by direct face-to-face care since medicine is based on four major principles: questioning, observation, palpation and auscultation. While the questioning can be carried out remotely using the telephone, and the observation (to a certain extent using videoconferencing), palpation and auscultation unfortunately cannot be carried out remotely.

Technology can certainly make healthcare more accessible, but this is still a long way from the virtual doctor who can completely replace in-person doctor visits. Due to distance, telemedicine can only provide very limited types of care. Direct care, such as blood analysis, X-rays, surgery or physiotherapy, is obviously not possible. Neither is care for chronic or complex diseases.

6 Available at: https://www.medicaline-sante.fr [Accessed October 10, 2016].

There are some questions about the sharing of responsibilities, ownership and data security and misuse of remote interpretation. The development of telemedicine must only be conducted with the aim of benefiting patients and their families, i.e. improving diagnosis, follow-ups, treatment time and quality of treatment. It is a tool in the service of patient care; it is a means and not an end. Its practice, procedures, technical installations and maintenance must take into account medical ethics.

## 11.2. Study of telemedicine in Africa

Telemedicine, as organized in Africa, consists of the practice of medical procedures *through* information and telecommunication technologies. In practice, it requires the participation of a medical professional. As a result, it has many benefits for patients, health professionals and the health system.

For patients, the benefits of telemedicine in Africa are mainly due to better quality of care. This means better access to care, especially for isolated or geographically remote patients, or those without a health professional in their vicinity.

It also results in better medical follow-up, especially for patients with chronic diseases[7]. Medical telemonitoring ensures continuous follow-up of the patient by the doctor, even during the interval between two consultations. The latter can then act early in case of a significant change in the patient's condition.

It is also a faster method of treatment, especially in emergency situations that require medical expertise (head injuries, stroke, etc.).

Finally, it involves a high level of security for medical procedures. Telemedicine is not "cheap" medicine or a second-best solution. On the contrary, it often increases the patient's survival chances, allowing for faster and more relevant care and referral within the healthcare system.

The benefits for health professionals are also notable and are similar to those of patients. Beyond this simple observation, telemedicine brings

7 For example, according to the Agence Anadolu (www.aa.com.tr), in Mali, telemedicine made it possible to coordinate a surgical humanitarian mission to treat cleft lip and palate (harelip) in 2012.

specific benefits to professionals. It promotes medical practice based on cooperation with other professionals, allowing everyone to better express their skills. It provides new possibilities for medical organizations, particularly in the field of continuing care (on-site and on-call duty)[8].

However, these benefits also imply constraints for health professionals such as changes in medical practice, managing the patient's "remote" relationship or the introduction of a new organizational scheme (Pierre 2015).

These aspects are obviously very important and must be the subject of sustained and adapted support in order to enable health professionals to master the practice of telemedicine.

It is important to point out that the health system as a whole also benefits from the development of telemedicine by providing better quality care in the situations mentioned above and by responding, at least partially, to some of the problems facing the country's health system, i.e. medical demographics, organization of the permanence of care and accessibility to care for all. It should also be mentioned that telemedicine contributes to the development of medical progress as well as other therapeutic or clinical innovations. Finally, telemedicine makes it possible to provide some services while controlling health expenses, through the efficient use of resources.

In general, remote medical practices via the Internet have been developing in recent years throughout the African continent. They make it possible to fill certain gaps in failing and inequitable health systems In addition to the inadequacy of health infrastructures and doctors, access to healthcare is poorly distributed on the continent with most technicians and specialists working in the capitals.

Thus, many doctors are often reluctant to go to work in rural areas of the country and telemedicine is a solution against their isolation. This technique offers them new skills through distance learning. In addition, additional income improves the ordinary life of these isolated physicians.

For patients in rural areas, medical examinations such as remote ultrasounds and electrocardiograms are available. In addition to saving time,

8 Available at: http://www.ameli.fr [Accessed October 31, 2016].

the low cost of telemedicine is also highlighted. According to the RAFT[9], several patients who received a remote ultrasound saved an average of 35,000 CFA francs (€53) per consultation on the transportation expense they would incur if they had gone to the health center where the requested services are available[10].

Medical desertification is not a new reality. Whether sectoral or geographical, the lack of health professionals in a given area represents a failure that needs to be compensated. To better organize the distribution of practitioners and promote access to care, the use of telemedicine is more than necessary.

Thus, telemedicine encompasses all medical practices whose main aim is to improve access to care through the use of new information and communication technologies. Its main objective is to improve the quality and efficiency of medical care by eliminating all constraints related to distance and geographical dispersion, thus enabling patients to receive appropriate care.

Today more than ever, telemedicine is at the heart of a revolution in the medical landscape of a continent in need of health professionals and real health structures. Africa is a continent that represents 13.8% of the world population but whose number of workers in the health sector is very low (they represent only 1.3% of health professionals in the world), according to 2013 World Health Statistics of the World Health Organization[11] and RAFT[12].

---

9 The RAFT (*Réseau d'Afrique francophone pour la télémédecine*), created in 2001 by hospitals and the University of Geneva, on the initiative of doctors and students from Mali, is also present in Cameroon. However, it is different from Genesis Telecare because it only provides distance learning. Genesis Telecare provides remote consultations (teleconsultation), remote diagnosis and remote expertise. Unlike RAFT, which only deals with doctors, Genesis Telecare is in direct contact with patients.

10 Available at: https://www.la-croix.com [Accessed November 2, 2016].

11 Available at: https://aa.com.tr [Accessed September 12, 2016].

12 Today, more than a thousand professionals participate in the RAFT, in about 20 African countries including Mali, Burkina Faso, Congo, Cameroon, etc. They follow continuing education courses, ask for help at a distance, to interpret an X-ray, an ultrasound image, an electrocardiogram plot, a photo of a skin lesion, or to obtain advice on how to manage a difficult case.

In summary, despite the positive characteristics of these developments, telemedicine raises many questions and merits maintaining permanent vigilance wherever it is practiced on the continent. It is a medical plus that must support the supply of care in Africa.

## 11.3. Specificity of Genesis Telecare in Cameroon

Genesis Telecare is a telemedicine method in Cameroon that provides access to quality healthcare through its telemedicine platform. It brings the patient closer to the doctor, reduces waiting times and maximizes the efficiency of medical consultations. It also provides rural areas with access to specialized examinations and services such as sustained medical monitoring of cardiovascular disease, difficult pregnancies or other early childhood care. The main medical service offered by telemedicine is teleconsultation, i.e. remote consultation between a doctor and a patient, the latter being able to be assisted by another health professional. There is also tele-expertise, which allows a medical professional to remotely seek the opinion of one or more medical professionals due to their specific training or skills, on the basis of medical information related to the management of a patient[13].

In addition, medical telemonitoring is an important telemedicine service that allows a medical professional to remotely interpret the data necessary for a patient's medical follow-up and, if necessary, to make decisions regarding the management of that patient. Data recording and transmission can be automated or performed by the patient himself or by a healthcare professional.

It is also worth mentioning teleassistance, which is a service that makes it possible for a health professional to provide remote medical assistance to another health professional during the performance of a procedure. Finally, telemedicine allows for an urgent medical response as part of regulatory activities.

It thus ensures easy access to specialist doctors without having to travel long distances. The practice offers an electronic medical record that is accessible to the most disadvantaged. Hence its success in its areas of

13 Available at: https://www.macsf-exerciceprofessionnel.fr [Accessed November 3, 2016].

operation. Some statistics show an attendance rate of more than three thousand people per month[14].

According to the World Health Organization, there is one doctor for every 10 thousand inhabitants in Cameroon. These doctors often work in large cities, but in the countryside, the supply of care is largely insufficient[15].

The services most commonly offered by telemedicine in Cameroon are cardiology, screening of cardiovascular diseases, ultrasound and electrocardiogram (ECG). The cost of consultations has been significantly reduced for patients. The patient who goes to any Genesis center in rural areas must pay 600 CFA francs (1 €) for a remote consultation by a general practitioner, 2,000 francs (3 €) if it is a specialist and 5,000 francs (7 €) for an examination such as an ultrasound or an electrocardiogram (Alajouanine and Grebot 2003).

Telemedicine is very useful for pregnant women who previously had to make regular trips to their attending physician for a pregnancy checkup. By contacting their doctor via the Internet, pregnant women can communicate freely without having to go to a health facility. Pregnancy checkups are improved, and risks are better controlled. In addition, a physical consultation is obviously more expensive than a teleconsultation even without transportation costs. The telemedicine system also allows medical personnel to make decisions more quickly, to better organize themselves and to advise and reassure their patients.

## 11.4. The Genesis Telecare approach in Cameroon

Genesis Telecare opened the first telemedicine center in Cameroon in April 2009[16]. The project is based on a team of general practitioners and specialists, nurses, engineers and technicians qualified in advanced technologies. Their collaboration allows real-time interaction between doctors and patients thanks to a highly secure computer system where each patient has a complete electronic medical record, regardless of where they live throughout Cameroon.

14 Available at: www.sinotables.com [Accessed November 30, 2016].

15 Available at: https://www.who.int [Accessed December 10, 2016].

16 Available at: www.icicemac.com [Accessed November 18, 2016].

In the countryside, the process as led by Genesis Telecare, based in Yaoundé, aims to "put technologies at the service of access to care and the quality and safety of care for all our fellow citizens"[17]. The aim is to make all the technologies that facilitate surveillance, diagnosis, expertise and even remote care available to citizens living in rural areas. To practice in this field, we do not need to be a doctor, although we must work in collaboration with doctors. In order to be recruited in a health facility of this type, being an information and communication technology professional is more preferable (Alajouanine and Grebot 2003).

The aim is to provide people living in remote areas access to doctors, including specialists. Most of the doctors involved in this project are based in Yaoundé, the capital of Cameroon, but some specialists consult from Switzerland, France or the United States.

To overcome the problem of frequent power outages in rural areas of Cameroon, generators are installed in each telemedicine center and some instruments use solar energy. For example, the Abong-Mbang[18] telemedicine center is equipped with computer equipment (computers equipped with devices such as cameras, headsets and microphones) and specific software adapted to telemedicine. It has a redundant power supply, a VSAT for Internet connection and sophisticated security systems. The servers are in different locations and protected highly secure hosting infrastructures. This protects the data and prevents any loss of information. Different equipment is used, including electronic electrocardiograms, electronic ultrasound scanners and electronic stethoscopes.

This center is managed by the hospital doctor and his staff who are all trained in the use of the equipment. When a patient goes to a telemedicine site, they are admitted by the district hospital staff. The doctor takes the parameters (weight, blood pressure, etc.) and transmits this information to the specialist at the Yaoundé technical center. The patient, assisted by the nurse, is then put in contact by videoconference with the specialist for

17 Available at: https://www.journalducameroun.com [Accessed October 4, 2016].

18 Genesis Telecare has already opened six telemedicine centers in Cameroon: Yaoundé (where the technical platform is located), Abong-Mbang District Hospital (in the east), Yagoua Regional Hospital (in the far north), SOSUCAM Health Center in Nkoteng (in the central region), Endom District Medical Center in Akonolinga (in the central region) and Nanga-Eboko.

a remote consultation. If necessary, electrocardiograms or ultrasounds are performed and sent to the specialist live or later on. The specialist interprets the examination and prescribes a medical prescription, which they refer to the doctor located at the telemedicine site (tele-diagnosis). The doctor located at the Abong-Mbang center requires the opinion of the specialist (tele-expertise).

Following success in Cameroon, telemedicine centers are being established in Gabon, Chad and the Democratic Republic of the Congo.

Telemedicine is at the heart of a revolution in the medical landscape of a continent in need of health professionals and real health structures.

Genesis Telecare makes it possible to carry out more precise diagnoses, interpret an X-ray, an ultrasound image and a photo of a skin lesion, or obtain advice on how to manage a difficult case. It is a large telemedicine network in Africa. To carry out its activities, Genesis Telecare has a satellite Internet connection and computer tools, usually powered by solar panels, to work on site.

## 11.5. Telemedicine and technological innovation in Africa

The African continent is becoming covered with antennas for cellular telephony. The Internet is spreading to small towns and governments are working to implement national strategies for e-health and telemedicine. Major humanitarian and medical development assistance programs, which lament the difficulty of coordinating their activities in the field, are investing in information and communication technologies. In response to the crisis caused by the lack of trained and active health professionals in the field, the World Health Organization is relying on telemedicine and distance learning tools.

For many medical activities, these tools make it possible to overcome distances and often deficient land transport routes, promoting collaboration between health professionals, improving the use of available resources, as well as the working conditions and efficiency of isolated health professionals, both for care activities and for health development and prevention activities.

As is often the case, and especially in medicine, the establishment of a relationship of trust is essential for effective exchange. However, in the case of telemedicine, distance, imperfect communication tools, organizational challenges, as well as socio-cultural differences can create misunderstandings or dysfunctions. It is therefore essential to support these remote exchanges through a human network of well-trained professionals who know each other, meet each other and share their experiences and knowledge. It is then that these differences and discrepancies can be used as sources of innovation and creativity.

With computer tools, doctors, who are based in the rural areas of the country, keep in touch with specialist doctors and their colleagues, continue training, seek advice when a case is difficult and decide whether a patient really should be referred to a big town or city. Expertise is displaced instead of patients.

These IT tools facilitate communication, collaboration and distance learning, to “de-isolate” health professionals and improve their working conditions and efficiency, and thereby the quality of patient care.

Participants are responsible for the operating costs of these tools and including the cost of Internet connection and technical support during the use of communication tools (ICT tools) in the performance of the medical procedure.

Before, doctors often used the telephone, and more recently, email to request specialized advice or refer patients to other colleagues. The recent emergence of videoconferencing capabilities via the Internet, combined with technological developments, opens up new possibilities for continuing medical advice requests and patient referrals under better conditions. Consultations, medical imaging (ultrasounds, scans, etc.), monitoring of vital parameters and even surgical procedures can be carried out remotely.

From a secure technical IT infrastructure, on which a platform offering collaborative services is installed, healthcare professionals perform remote medical services such as teleconsultation, teleassistance, tele-expertise, telemonitoring, medical regulation or multidisciplinary consultation meetings.

For health professionals, telemedicine provides technical solutions for collaboration, cooperation or supervision in the practice of medicine. The experience of tele-expertise, for example, shows that it facilitates the prioritization of patients in medical management. For patients, it facilitates the flow of care and access to specialized medical resources.

Outside South Africa and a few other rare countries, the African landscape is marked by insufficient medical coverage in terms of both quantity and quality. It should also be noted that there is insufficient investment in quality health infrastructure and equipment, and a relatively low level of education with respect to health problems. The arrival of telemedicine in this panorama can be a structuring element, making it possible to meet the challenges of access to quality specialized care.

The use of telemedicine is leading to the rethinking of health system organization. With this tool, the notions of distance and time are put into perspective. In this way, the expert medical resource and consequently the infrastructure and equipment can be concentrated within the framework of spatial planning adapted to cultural and political socio-economic constraints.

To support a medical project, it is necessary to find technical solutions in terms of appropriate communication infrastructures, a sufficiently ergonomic collaborative platform, high-performance connected objects (camera, medical equipment such as digital stethoscope, digital dermatoscope, ultrasound and ECG device). This is one of the reasons why investment in digital communication infrastructure in Africa has the potential to improve the health and well-being of populations.

The aim of telemedicine is to improve patient care by providing access to a network of specialists covering all medical and surgical specialities for careers in the field.

From location A, a local doctor can discuss the case of an injured person with his colleague in location B. Digital technology allows them to exchange with surgeon B and send them photos of their patient in preparation for a possible operation. With a single click, practitioners can send a medical image with their tablet or smartphone to ask a colleague for an opinion. Ultrasound and cardiological teleconsultation have proved their worth. They are developing throughout the continent and are helping to avoid unnecessary and costly medical evacuations.

However, telemedicine does not replace the face-to-face relationship between the doctor and the patient, nor does it replace the clinical examination. It does, however, effectively compensate for the shortage of specialists and infrastructure deficiencies, making accessible today what was not accessible yesterday.

Genesis Telecare has more than a hundred specialists in its database. In practice, the field doctor who needs an opinion sends a summary of the clinical examination via the Internet, with an X-ray or photo (anonymized) and, within a few hours, receives a response. The latter must not be out of line with local reality: it would be pointless, for example, to ask for a scan for a patient in the middle of the bushlands! Moreover, the referring practitioners are chosen according to their experience in the field.

## 11.6. Conclusion

Telemedicine is a form of remote medical practice using information and communication technologies. It connects one or more health professionals with each other or with a patient, and among them includes a medical professional and, where appropriate, other professionals (psychologists, sociologists, etc.) providing care to the patient.

It makes it possible to establish a diagnosis, to ensure, for a patient at risk, a follow-up for preventive purposes or post-treatment follow-up, to request specialized advice, to prepare a treatment decision, prescribe treatment, perform services or procedures or monitor the condition of patients.

Telemedicine is therefore primarily a medical practice, not a technological system.

To propose innovative solutions to the problems of access to healthcare in Cameroon and Africa, Genesis Telecare opened its very first telemedicine center in the country. The main objective presented was to provide patients with “quick and effective access” to general practitioners and specialists, using telemedicine technologies. It is about combining cutting-edge technologies with the expertise of qualified physicians in a real-time and personalized interaction from a highly secure computer system.

Due to technological developments that offer new opportunities in many African countries, we are witnessing the development of telemedicine to support patients, their families and medical and paramedical professionals.

## 11.7. References


Alajouanine, G. and Grebot, E. (2003). *Les nouvelles technologies de l'information au service de la santé en Afrique*. PUF, Paris.

Amenta, F. (2005). *Le Centro Internazionale Radio Medico (C.I.R.M.): 70 ans d'experience romaine*. C.I.R.M., Rome.

Bonjawo, J. (2001). *Révolution numérique dans les pays en développement, l'exemple africain*. Dunod, Paris.

Bonjawo, J. (2002). *Internet, une chance pour l'Afrique*. Karthala, Paris.

Lamine, S. (2015). *La télésanté. Impact et avenir en Afrique de l'Ouest*. Éditions universitaires européennes, Sarrebruck.

Pierre, S. (2015). *Télémédecine : enjeux et pratiques*. Le Coudrier, Mont-Saint-Guibert.

Severino, J.-M. and Ray, O. (2010). *Le Temps de l'Afrique*. Odile Jacob, Paris.


## 11.8. Webography


https://www.medicaline-sante.fr [Accessed 10 October 2016].

https://www.ameli.fr/professionnels-de-sante/medecins [Accessed 31 October 2016].

https://www.la-croix.com [Accessed 2 November 2016].

https://aa.com.tr/fr/sante [Accessed 12 September 2016].

https://www.macsf-exerciceprofessionnel.fr [Accessed 3 November 2016].

https://www.sinotables.com [Accessed 30 November 2016].

https://www.who.int [Accessed 10 December 2016].

www.icicemac.com [Accessed 18 November 2016].

https://www.journalducameroun.com [Accessed 4 October 2016].

# List of Authors

Ettien Franck Stéphane ADOU
Félix Houphouët-Boigny University
Abidjan
Côte d'Ivoire

Julien ATCHOUA
CERCOM
Félix Houphouët-Boigny University
Abidjan
Côte d'Ivoire

Insa BADJI
University of Thiès
Senegal

Marcel BAGARÉ
Norbert Zongo University
Koudougou
Burkina Faso

Jean-Jacques BOGUI
CERCOM
Félix Houphouët-Boigny University
Abidjan
Côte d'Ivoire

Moustapha CISSÉ
University of Thiès
Senegal

Évariste DAKOURÉ
Aube Nouvelle University
Ouagadougou
Burkina Faso

Mouhamadou S. DIALLO
University of Strasbourg
France

Saikou DIALLO
VMASC
Old Dominion University
Norfolk
United States

Oulaï Honoré KAHI
Alassane Ouattara University
Bouaké
Côte d'Ivoire

Olivier Koffi KOFFI
Félix Houphouët-Boigny University
Abidjan
Côte d'Ivoire

Mahesse Stéphanie KOLÉ
Félix Houphouët-Boigny University
Abidjan
Côte d'Ivoire

Raymond Kouassi KRA
Félix Houphouët-Boigny University
Abidjan
Côte d'Ivoire

Alain François LOUKOU
Alassane Ouattara University
Bouaké
Côte d'Ivoire

Gbati NAPO
University of Lomé
Togo

Marie NDIAYE
Assane Seck University
Ziguinchor
Senegal

Nathalie PINÈDE
MICA
Bordeaux Montaigne University
France

Ousmane SALL
University of Thiès
Senegal

Abdoulaye SY
University of Thiès
Senegal

Gilbert TOPPÉ
Alassane Ouattara University
Bouaké
Côte d'Ivoire

# Index

## A, B, C



## D, E, F

## G, H, I



## M, P, R



## S, T, V, W

Other titles from

ISTE

in

Systems and Industrial Engineering – Robotics

## 2019

ANDRÉ Jean-Claude
*Industry 4.0: Paradoxes and Conflicts*

BENSALAH Mounir, ELOUADI Abdelmajid, MHARZI Hassan
*Railway Information Modeling RIM: The Track to Rail Modernization*

BLUA Philippe, YALAOU Farouk, AMODEO Lionel, DE BLOCK Michaël, LAPLANCHE David
*Hospital Logistics and e-Management: Digital Transition and Revolution*

BRIFFAUT Jean-Pierre
*From Complexity in the Natural Sciences to Complexity in Operations Management Systems*
*(Systems of Systems Complexity Set – Volume 1)*

BUDINGER Marc, HAZYUK Ion, COÏC Clément
*Multi-Physics Modeling of Technological Systems*

FLAUS Jean-Marie
*Cybersecurity of Industrial Systems*

JAULIN Luc
*Mobile Robotics – Second Edition Revised and Updated*

KUMAR Kaushik, DAVIM Paulo J.
*Optimization for Engineering Problems*

TRIGEASSOU Jean-Claude, MAAMRI Nezha
*Analysis, Modeling and Stability of Fractional Order Differential Systems 1: The Infinite State Approach*
*Analysis, Modeling and Stability of Fractional Order Differential Systems 2: The Infinite State Approach*

VANDERHAEGEN Frédéric, MAAOUI Choubeila, SALLAK Mohamed, BERDJAG Denis
*Automation Challenges of Socio-technical Systems*

## 2018

BERRAH Lamia, CLIVILLÉ Vincent, FOULLOY Laurent
*Industrial Objectives and Industrial Performance: Concepts and Fuzzy Handling*

GONZALEZ-FELIU Jesus
*Sustainable Urban Logistics: Planning and Evaluation*

GROUS Ammar
*Applied Mechanical Design*

LEROY Alain
*Production Availability and Reliability: Use in the Oil and Gas Industry*

MARÉ Jean-Charles
*Aerospace Actuators 3: European Commercial Aircraft and Tiltrotor Aircraft*

MAXA Jean-Aimé, BEN MAHMOUD Mohamed Slim, LARRIEU Nicolas
*Model-driven Development for Embedded Software: Application to Communications for Drone Swarm*

MBIHI Jean
*Analog Automation and Digital Feedback Control Techniques*
*Advanced Techniques and Technology of Computer-Aided Feedback Control*

MORANA Joëlle
*Logistics*

SIMON Christophe, WEBER Philippe, SALLAK Mohamed
*Data Uncertainty and Important Measures*
*(Systems Dependability Assessment Set – Volume 3)*

TANIGUCHI Eiichi, THOMPSON Russell G.
*City Logistics 1: New Opportunities and Challenges*
*City Logistics 2: Modeling and Planning Initiatives*
*City Logistics 3: Towards Sustainable and Liveable Cities*

ZELM Martin, JAEKEL Frank-Walter, DOUMEINGTS Guy, WOLLSCHLAEGER Martin
*Enterprise Interoperability: Smart Services and Business Impact of Enterprise Interoperability*

## 2017

ANDRÉ Jean-Claude
*From Additive Manufacturing to 3D/4D Printing 1: From Concepts to Achievements*
*From Additive Manufacturing to 3D/4D Printing 2: Current Techniques, Improvements and their Limitations*
*From Additive Manufacturing to 3D/4D Printing 3: Breakthrough Innovations: Programmable Material, 4D Printing and Bio-printing*

ARCHIMÈDE Bernard, VALLESPIR Bruno
*Enterprise Interoperability: INTEROP-PGSO Vision*

CAMMAN Christelle, FIORE Claude, LIVOLSI Laurent, QUERRO Pascal
*Supply Chain Management and Business Performance: The VASC Model*

FEYEL Philippe
*Robust Control, Optimization with Metaheuristics*

MARÉ Jean-Charles
*Aerospace Actuators 2: Signal-by-Wire and Power-by-Wire*

POPESCU Dumitru, AMIRA Gharbi, STEFANOIU Dan, BORNE Pierre
*Process Control Design for Industrial Applications*

RÉVEILLAC Jean-Michel
*Modeling and Simulation of Logistics Flows 1: Theory and Fundamentals*
*Modeling and Simulation of Logistics Flows 2: Dashboards, Traffic Planning and Management*
*Modeling and Simulation of Logistics Flows 3: Discrete and Continuous Flows in 2D/3D*

## 2016

ANDRÉ Michel, SAMARAS Zissis
*Energy and Environment*
*(Research for Innovative Transports Set - Volume 1)*

AUBRY Jean-François, BRINZEI Nicolae, MAZOUNI Mohammed-Habib
*Systems Dependability Assessment: Benefits of Petri Net Models (Systems Dependability Assessment Set - Volume 1)*

BLANQUART Corinne, CLAUSEN Uwe, JACOB Bernard
*Towards Innovative Freight and Logistics (Research for Innovative Transports Set - Volume 2)*

COHEN Simon, YANNIS George
*Traffic Management (Research for Innovative Transports Set - Volume 3)*

MARÉ Jean-Charles
*Aerospace Actuators 1: Needs, Reliability and Hydraulic Power Solutions*

REZG Nidhal, HAJEJ Zied, BOSCHIAN-CAMPANER Valerio
*Production and Maintenance Optimization Problems: Logistic Constraints and Leasing Warranty Services*

TORRENTI Jean-Michel, LA TORRE Francesca
*Materials and Infrastructures 1 (Research for Innovative Transports Set - Volume 5A)*
*Materials and Infrastructures 2 (Research for Innovative Transports Set - Volume 5B)*

WEBER Philippe, SIMON Christophe
*Benefits of Bayesian Network Models*
*(Systems Dependability Assessment Set – Volume 2)*

YANNIS George, COHEN Simon
*Traffic Safety (Research for Innovative Transports Set - Volume 4)*

## 2015

AUBRY Jean-François, BRINZEI Nicolae
*Systems Dependability Assessment: Modeling with Graphs and Finite State Automata*

BOULANGER Jean-Louis
*CENELEC 50128 and IEC 62279 Standards*

BRIFFAUT Jean-Pierre
*E-Enabled Operations Management*

MISSIKOFF Michele, CANDUCCI Massimo, MAIDEN Neil
*Enterprise Innovation*

## 2014

CHETTO Maryline
*Real-time Systems Scheduling*
*Volume 1 – Fundamentals*
*Volume 2 – Focuses*

DAVIM J. Paulo
*Machinability of Advanced Materials*

ESTAMPE Dominique
*Supply Chain Performance and Evaluation Models*

FAVRE Bernard
*Introduction to Sustainable Transports*

GAUTHIER Michaël, ANDREFF Nicolas, DOMBRE Etienne
*Intracorporeal Robotics: From Milliscale to Nanoscale*

MICOUIN Patrice
*Model Based Systems Engineering: Fundamentals and Methods*

MILLOT Patrick
*Designing Human–Machine Cooperation Systems*

NI Zhenjiang, PACORET Céline, BENOSMAN Ryad, RÉGNIER Stéphane
*Haptic Feedback Teleoperation of Optical Tweezers*

OUSTALOUP Alain
*Diversity and Non-integer Differentiation for System Dynamics*

REZG Nidhal, DELLAGI Sofien, KHATAD Abdelhakim
*Joint Optimization of Maintenance and Production Policies*

STEFANOIU Dan, BORNE Pierre, POPESCU Dumitru, FILIP Florin Gh.,
EL KAMEL Abdelkader
*Optimization in Engineering Sciences: Metaheuristics, Stochastic Methods and Decision Support*

## 2013

ALAZARD Daniel
*Reverse Engineering in Control Design*

ARIOUI Hichem, NEHAOUA Lamri
*Driving Simulation*

CHADLI Mohammed, COPPIER Hervé
*Command-control for Real-time Systems*

DAAFOUZ Jamal, TARBOURIECH Sophie, SIGALOTTI Mario
*Hybrid Systems with Constraints*

FEYEL Philippe
*Loop-shaping Robust Control*

FLAUS Jean-Marie
*Risk Analysis: Socio-technical and Industrial Systems*

FRIBOURG Laurent, SOULAT Romain
*Control of Switching Systems by Invariance Analysis: Application to Power Electronics*

GROSSARD Mathieu, REGNIER Stéphane, CHAILLET Nicolas
*Flexible Robotics: Applications to Multiscale Manipulations*

GRUNN Emmanuel, PHAM Anh Tuan
*Modeling of Complex Systems: Application to Aeronautical Dynamics*

HABIB Maki K., DAVIM J. Paulo
*Interdisciplinary Mechatronics: Engineering Science and Research Development*

HAMMADI Slim, KSOURI Mekki
*Multimodal Transport Systems*

JARBOUI Bassem, SIARRY Patrick, TEGHEM Jacques
*Metaheuristics for Production Scheduling*

KIRILLOV Oleg N., PELINOVSKY Dmitry E.
*Nonlinear Physical Systems*

LE Vu Tuan Hieu, STOICA Cristina, ALAMO Teodoro,
CAMACHO Eduardo F., DUMUR Didier
*Zonotopes: From Guaranteed State-estimation to Control*

MACHADO Carolina, DAVIM J. Paulo
*Management and Engineering Innovation*

MORANA Joëlle
*Sustainable Supply Chain Management*

SANDOU Guillaume
*Metaheuristic Optimization for the Design of Automatic Control Laws*

STOICAN Florin, OLARU Sorin
*Set-theoretic Fault Detection in Multisensor Systems*

## 2012

AÏT-KADI Daoud, CHOUINARD Marc, MARCOTTE Suzanne, RIOPEL Diane
*Sustainable Reverse Logistics Network: Engineering and Management*

BORNE Pierre, POPESCU Dumitru, FILIP Florin G., STEFANOIU Dan
*Optimization in Engineering Sciences: Exact Methods*

CHADLI Mohammed, BORNE Pierre
*Multiple Models Approach in Automation: Takagi-Sugeno Fuzzy Systems*

DAVIM J. Paulo
*Lasers in Manufacturing*

DECLERCK Philippe
*Discrete Event Systems in Dioid Algebra and Conventional Algebra*

DOUMIATI Moustapha, CHARARA Ali, VICTORINO Alessandro, LECHNER Daniel
*Vehicle Dynamics Estimation using Kalman Filtering: Experimental Validation*

GUERRERO José A, LOZANO Rogelio
*Flight Formation Control*

HAMMADI Slim, KSOURI Mekki
*Advanced Mobility and Transport Engineering*

MAILLARD Pierre
*Competitive Quality Strategies*

MATTA Nada, VANDENBOOMGAERDE Yves, ARLAT Jean
*Supervision and Safety of Complex Systems*

POLER Raul *et al.*
*Intelligent Non-hierarchical Manufacturing Networks*

TROCCAZ Jocelyne
*Medical Robotics*

YALAOUI Alice, CHEHADE Hicham, YALAOUI Farouk, AMODEO Lionel
*Optimization of Logistics*

ZELM Martin *et al.*
*Enterprise Interoperability –I-EASA12 Proceedings*

## 2011

CANTOT Pascal, LUZEAUX Dominique
*Simulation and Modeling of Systems of Systems*

DAVIM J. Paulo
*Mechatronics*

DAVIM J. Paulo
*Wood Machining*

GROUS Ammar
*Applied Metrology for Manufacturing Engineering*

KOLSKI Christophe
*Human–Computer Interactions in Transport*

LUZEAUX Dominique, RUAULT Jean-René, WIPPLER Jean-Luc
*Complex Systems and Systems of Systems Engineering*

ZELM Martin, *et al.*
*Enterprise Interoperability: IWEI2011 Proceedings*

## 2010

BOTTA-GENOULAZ Valérie, CAMPAGNE Jean-Pierre, LLERENA Daniel, PELLEGRIN Claude
*Supply Chain Performance / Collaboration, Alignement and Coordination*

BOURLÈS Henri, GODFREY K.C. Kwan
*Linear Systems*

BOURRIÈRES Jean-Paul
*Proceedings of CEISIE'09*

CHAILLET Nicolas, REGNIER Stéphane
*Microrobotics for Micromanipulation*

DAVIM J. Paulo
*Sustainable Manufacturing*

GIORDANO Max, MATHIEU Luc, VILLENEUVE François
*Product Life-Cycle Management / Geometric Variations*

LOZANO Rogelio
*Unmanned Aerial Vehicles / Embedded Control*

LUZEAUX Dominique, RUAULT Jean-René
*Systems of Systems*

VILLENEUVE François, MATHIEU Luc
*Geometric Tolerancing of Products*

## 2009

DIAZ Michel
*Petri Nets / Fundamental Models, Verification and Applications*

OZEL Tugrul, DAVIM J. Paulo
*Intelligent Machining*

PITRAT Jacques
*Artificial Beings*

## 2008

ARTIGUES Christian, DEMASSEY Sophie, NERON Emmanuel
*Resources–Constrained Project Scheduling*

BILLAUT Jean-Charles, MOUKRIM Aziz, SANLAVILLE Eric
*Flexibility and Robustness in Scheduling*

DOCHAIN Denis
*Bioprocess Control*

LOPEZ Pierre, ROUBELLAT François
*Production Scheduling*

THIERRY Caroline, THOMAS André, BEL Gérard
*Supply Chain Simulation and Management*

## 2007

DE LARMINAT Philippe
*Analysis and Control of Linear Systems*

DOMBRE Etienne, KHALIL Wisama
*Robot Manipulators*

LAMNABHI Françoise *et al.*
*Taming Heterogeneity and Complexity of Embedded Control*

LIMNIOS Nikolaos
*Fault Trees*

## 2006

FRENCH COLLEGE OF METROLOGY
*Metrology in Industry*

NAJIM Kaddour
*Control of Continuous Linear Systems*